D0873840

ENGINEERING FORMULAS

INTERACTIVE

Conversions, Definitions, and Tables

BY
Frank Sims

INDUSTRIAL PRESS INC.

Library of Congress Cataloging in Publication Data

Sims, Frank, 1925-
    Engineering formulas : conversions, definitions, and tables / by
Frank Sims. – Interactive ed.
    416p. 33x42cm.
    Includes index.
    ISBN 0-8311-3087-3 (alk. Paper)
    1   Engineering mathematics—Formulae. 2. Engineering—Tables.
3.   Materials—Tables.   I. Title.
TA332.S55   1999
620'.002'12—dc21   98-49740
                                           CIP

Industrial Press, Inc.
200 Madison Avenue
New York, NY 10016-4078

First Interactive Edition, January 1999

Sponsoring Editor: John Carleo
Project Editor: Sheryl A. Levart
Book & Cover Design: Janet Romano
Software: Intellipro, Inc.

Printed in the United States of America

10 9 8 7 6 5 4 3

PREFACE

Early in my engineering career I discovered the need for a reference book that would be comprehensive and easy to use. The standard engineering handbooks left me frustrated—not only because they often explained theories rather than providing the formulas or facts that I needed, but also because the search for information on a given topic would require looking through several books before a satisfactory answer could be found. Why not, I thought, compile an all inclusive compendium of the basic data and information that I actually needed and used, and assemble it in alphabetical order to make it accessible? This book, which grew from the "manuscript" that I began more than thirty years ago, is the result of that inspiration. It includes formulas for engineering subjects, over one hundred geometric figures, definitions, structural sections, properties of materials, unit conversions, and much more that will be of use to engineers, designers, and students.

Here are a few a tips for getting the most out of *Engineering Formulas*:

Formula sections that are lengthy and may run for several pages are identified by *boldface capital letters*. As an example, the heading **BEAM LOAD FORMULAS** designates a long formula section that continues until terminated by a *horizontal line* across the page.

Where several terms have the same meaning: An *asterisk* following a reference indicates that the formula for the topic will be found under another listing in the same section. In the ELECTROSTATICS section, for example, "**capacitivity**: dielectric constant *" indicates that the formula for "capacitivity" will be found in the electrostatics section under the heading "dielectric constant."

Unit conversions are found alphabetically throughout the book. Rather than repeat entries of same (or equal) value, multiple entries are cross referenced and the target reference is indicated by *capital letters*. For instance "**millipascal-second**: CENTIPOISE" indicates that the unit conversions for "millipascal-second" will be found under "centipoise."

Terminology entries of the same meaning are cross referenced to one selected form. Consider the following examples:

dynamic viscosity: VISCOSITY, ABSOLUTE
progression, geometric: GEOMETRIC PROGRESSION

In these instances, information about the topics identified with *bold letters* will be found in the section designated by *capital letters*.

It has been my intent to include as much information as possible. However, should you discover that formulas, units, or conversions that you find useful are not included, I would appreciate hearing from you. Any omissions can hopefully be rectified in future editions.

Frank Sims, August, 1996

Abbreviations and Unit Symbols

α	atto (10^{-18})
A	ampere
Å	angstrom
acfm	actual cubic feet per minute
A-h	ampere-hour
apoth	apothecary
A-s	ampere-second
atm	atmosphere
avdp	avoirdupois
B	bel
bbl	barrel
BHN	Brinell Hardness Number
bhp	brake horsepower
Btu	British thermal unit
BWG	Birmingham Wire Gauge
c	centi (10^{-2})
c	cycle
C	Celsius
C	coulomb
C_P	specific heat (constant pressure)
C_V	specific heat (constant volume)

°C	degree Celsius (temperature)
cal	calorie
cd	candela (preferred to candle)
cfm	cubic feet per minute
cfs	cubic feet per second
cg	center of gravity
cg	centigram
cgs	centimeter-gram-second
cl	centiliter
cm	centimeter
cmil	circular mil
cP	centipoise
cSt	centistoke
d	day
d	deci (10^{-1})
da	deka (10)
dag	dekagram
dal	dekaliter
dam	dekameter
das	dekastere
dB	decibel
deg	degree
dg	decigram
dl	deciliter
dm	decimeter
dr	dram
ds	decistere
dwt	pennyweight
dyn	dyne
eff	efficiency
emf	electromotive force
eV	electronvolt
f	femto (10^{-15})
F	Fahrenheit

F	farad
°F	degree Fahrenheit (temperature)
fpm	foot per minute
fps	foot per second
ft	foot, feet
ft-lb	foot-pound
ft/min	foot per minute
ft/s	foot per second
ft^3/min	cubic foot per minute
ft^3/s	cubic foot per second
g	gram
g	gravitational acceleration constant
G	gauss, pl. gauss
G	giga (10^9)
gal	gallon
gal/min	gallon per minute
Gb	gilbert
gpm	gallon per minute
gr	grain
h	hecto (10^2)
h or hr	hour
H	henry, pl. henrys
ha	hectare
hg	hectogram
hl	hectoliter
hm	hectometer
hp	horsepower
hw	hectowatt
hwt	hundredweight
Hz	hertz
in.	inch
in-lb	inch-pound

J	joule
k	kilo (10^3)
K	degree kelvin (temperature)
K	kelvin
kc	kilocycle
kg	kilogram
kJ	kilojoule
kl	kiloliter
km	kilometer
kn	knot
kPa	kilopascal
kV	kilovolt
kVA	kilovolt-ampere
kVAR	kilovar
kW	kilowatt
kWh	kilowatthour
l	liter
L	lambert
lb	pound
lm	lumen
lx	lux
m	meter
m	milli (10^{-3})
M	mega (10^6)
mA	milliampere
mbar	millibar
Mbar	megabar
Mc	megacycle
mcm	1000 circular mils
mg	milligram
mH	millihenry
MHz	megahertz

mi	mile
min	minute
mks	meter-kilogram-second
ml	milliliter
mL	millilambert
mm	millimeter
Mm	megameter
mol	mole
mPa	millipascal
MPa	megapascal
mph	milliphot
mph	miles per hour
mV	millivolt
MV	megavolt
MW	megawatt
MW	molecular weight
Mx	maxwell
n	nano (10^{-9})
N	newton
N_{Nu}	Nusselt number
N_P	Nepcr
N_R	Reynold's number
naut	nautical
nm	nanometer
N-m	newton-meter
Oe	oersted
oz	ounce (avdp)
p	pico (10^{-12})
P	poise
Pa	pascal
pdl	poundal
pF	picofarad
ph	phot
pm	picometer

ppm	parts per million
psf	pound per square foot
psi	pound per square inch
psia	pound per square inch absolute pressure
psig	pound per square inch gauge pressure
pt	pint
pwt	pennyweight
qt	quart
R	Rankine
°R	degree Rankine (temperature)
rad	radian
rev	revolution
rev/min	revolution per minute
rev/s	revolution per second
rpm	revolution per minute
rps	revolution per second
s or sec	second (time)
sp gr	specific gravity
sq	square
sr	steradian
SSF	Saybolt Second Furol
SSU	Saybolt Second Universal
St	stoke
stat	statute
SUS	Saybolt Second Universal
T	tera (10^{12})
tbsp	tablespoon
tsp	teaspoon
V	volt
VA or V-A	voltampere
W	watt
Wb	weber

Wh or W-h	watthour
Ws or W-s	wattsecond
yd	yard
μ	micro (10^{-6})
μA	microampere
μbar	microbar
μC	microcoulomb
μF	microfarad
μg	microgram
μH	microhenry
μl	microliter
μm	micrometer
μP	micropoise
μV	microvolt

Index of Interactive Equations on the CD-ROM

A

A: argon (element)

abampere =

    10 A   2.998 E + 10 statamperes
    1.036 E − 04 faraday/s

abampere-turn =

    10 A-turns
    12.566 Gb

abcoulomb =

    0.002778 A-h   1.036 E − 04 faraday
    10 C   2.998 E + 10 statcoulombs
    6.2425 E + 19 electronic charges

abfarad =

    1 E + 09 F   8.988 E + 20 statfarads
    1 E + 15 μF

abhenry =

    1 E − 09 H   1.111 E − 21 stathenry
    1 E − 06 mH

1

abmho =

1000 megmhos	1 E + 15 micromhos
1 E + 09 mhos	8.988 E + 20 statmhos

abmho/centimeter; abmho/cubic centimeter: metric unit for electrical conductivity of a material, based on cross-sectional area of 1 cm^2, in abmhos per centimeter length.

abmho/centimeter = 1 E + 09 mhos/cm

abmho/cubic centimeter: ABMHO/CENTIMETER

abohm =

1 E − 15 megohm	1 E − 09 ohm
0.001 microhm	1.113 E − 21 statohm

abohm-centimeter; abohm-centimeter cube: metric unit for resistivity of a material, based on a cross-sectional area of 1 cm^2, in abohms per centimeter length.

abohm-centimeter = 1 E − 09 ohm-cm

absolute density: DENSITY, ABSOLUTE

absolute humidity: HUMIDITY, ABSOLUTE

absolute pressure: atmospheric pressure plus gauge pressure, pressure above a perfect vacuum, designated as psia.

$$\begin{aligned} \text{psia} &= \text{gauge pressure} + 14.696 \\ &= 0.433[\text{ft (water) gauge} + 33.898] \\ &= 0.491[\text{in. (mercury) gauge} + 29.92] \\ &= 0.036[\text{in. (water) gauge} + 406.77] \end{aligned}$$

absolute system of units: English system using the units of force in poundals and mass in pounds.

absolute temperature degrees Celsius: same as Kelvin (K).

$$°C, \text{absolute} = K = °C + 273.16$$

absolute temperature degrees Fahrenheit: same as degrees Rankine (°R).

$$°F, \text{absolute} = °R = °F + 459.69$$

absolute viscosities of various gases: GAS FLOW
absolute viscosities of various liquids: FLUID FLOW
absolute viscosity: VISCOSITY, ABSOLUTE
absolute zero temperature =

$-273.16°C$	0 K
$-459.69°F$	$0°R$

absorptivity: HEAT TRANSFER
abvolt =

0.01 μV	3.336 E-11 statvolt
1 E-05 mV	1 E-08 V

Ac: actinium (element)
acceleration: the continuous increase of velocity of a body in motion, as in linear motion or rotation.
acceleration force: force applied to accelerate a body.
acceleration of gravity: the continuous increase of the velocity of a free-falling body due to its own weight, designated by the letter g.
acceleration of gravity (g) =

980.665 cm/s^2
32.174 ft/s^2

acre =

40.47 ares	0.004047 km^2
4046.87 centares	4046.87 m^2
4.0468 E$+07$ cm^2	0.0015625 mi^2
40.469 dam^2	160 sq. perches
43560 ft^2	160 sq. rods
0.40469 ha	4840 yd^2
6.2726 E$+06$ in.^{2}	

acre-foot =

43560 ft^3	1233.49 m^3
325,851 gal (liquid)	1613.3 yd^3

acre-foot/hour =

> 12.1 ft^3/s
> 5431 gal/min

acre-inch =

> 3630 ft^3 102.79 m^3
> 27,154 gal (liquid)

acre-inch/hour =

> 1.008 ft^3/s
> 452.6 gal/min

acute angle: an angle of less than 90°.
adiabatic compression of a gas: THERMODYNAMICS
adiabatic expansion: THERMODYNAMICS
adiabatic exponent: THERMODYNAMICS
adiabatic process: THERMODYNAMICS
Ag: silver (element)
AIR:

> **adiabatic exponent (k):** 1.4
> **boiling point:** −317°F
> **composition, percent by volume:**
>
> > nitrogen: 78.03   carbon dioxide: 0.035
> > oxygen: 20.99   hydrogen: 0.010
> > argon: 0.93   others: 0.005
>
> **composition, percent by weight:**
>
> > nitrogen: 75.1   argon: 1.2
> > oxygen: 23.1   others: 0.6
>
> **density:** specific weight *
> **gas constant (R):** 53.35 ft-lb/lb/°R
> **molecular weight:** 28.9752
> **specific gravity:** 1.000

AIR — *continued*

specific heat at 70°F:

$C_P = 0.239$ Btu/lb/°F, at constant pressure
$C_V = 0.171$ Btu/lb/°F, at constant volume

specific volume at atmospheric pressure:

at 32°F = 12.389 ft^3/lb
at 70°F = 13.333 ft^3/lb

specific weight at atmospheric pressure:

at 32°F = 0.08072 lb/ft^3
at 70°F = 0.075 lb/ft^3

viscosity: 0.018 cP

air compression formulas: AIR/GAS COMPRESSION FORMULAS
AIR FLOW FORMULAS: where

a = height of duct, in.
A = area, in.^{2}
acfm = actual ft^3/min
b = width of duct, in.
C_B = coefficient of duct bend
C_F = coefficient of duct fitting
d = density, slugs/ft^3
D = inside diameter of duct or pipe, in.
f = friction factor
g = gravitational acceleration = 32.174 ft/sec^2
h = head, in. (water)
h_F = head loss, friction, in. (water)
h_L = head loss, in. (water)
h_P = head, potential energy, in. (water)
h_S = head, static pressure, in. (water)
h_T = head, total pressure, in. (water)
h_V = head, velocity pressure, in. (water)
L = length of duct or pipe, ft
N_R = Reynold's number
p = gauge pressure, psig

AIR FLOW FORMULAS — *continued*

p_1 = initial pressure, psig
p_2 = final pressure, psig
P = absolute pressure, psia
P_1 = initial pressure, psia
P_2 = final pressure, psia
Q_A = actual flow rate, acfm
Q_S = standard flow rate, scfm
Q_W = flow rate by weight, lb/min
R = gas constant, ft-lb/lb/°R
s = perimeter of duct, in.
scfm = standard ft^3/min
t = temperature, °F
T = absolute temperature, °R
v = volume, ft^3
v_1 = initial volume, ft^3
v_2 = final volume, ft^3
V = velocity of air, ft/min
w = specific weight of air, lb/ft^3
W = weight of air, lb
μ = absolute viscosity, cP
v = kinematic viscosity, cSt

absolute pressure, psia:

$$P = p + 14.696$$
$$= 0.433[\text{ft (water) gauge} + 33.898]$$
$$= 0.491[\text{in. (mercury) gauge} + 29.92]$$
$$= 0.036[\text{in. (water) gauge} + 406.77]$$

absolute temperature, °R:

$$T = t + 459.69$$

absolute viscosity: VISCOSITY, ABSOLUTE
absolute viscosity (μ) of air = 0.018 cP
actual cubic feet per minute, acfm (Q_A): actual flow rate by volume.
area: cross-sectional area of duct or pipe,

rectangular duct, $A = ab$
round duct or pipe, $A = 0.7854D^2$

AIR FLOW FORMULAS — *continued*

Bernoulli's equation: conservation of energy; total energy is constant in the steady flow of an incompressible gas, whereby potential energy + pressure energy + velocity energy equals a constant.

$$h_P + h_S + h_V = \text{constant}$$

or between points 1 and 2,

$$h_{P1} + h_{S1} + h_{V1} = h_{P2} + h_{S2} + h_{V2}$$

coefficient (C_B) of 90° duct bend:

for conversion to head loss (h_L),
rectangular duct, where R_B = duct width/bend radius,

R_B	0.50	0.67	0.80	1.0	1.3	2.0	4.0
C_B	0.08	0.09	0.12	0.2	0.3	0.9	1.1

round duct, where R_B = duct diameter/bend radius,

R_B	0.50	0.67	0.80	1.0	1.3	2.0	4.0
C_B	0.15	0.17	0.20	0.25	0.40	0.75	0.80

coefficient (C_F) of duct fitting:

for conversion to head loss (h_L),
contraction to smaller size, $C_F =$
15° taper: 0.1   60° taper: 0.30
30° taper: 0.15

enlargement to larger size, $C_F =$
5° taper: 0.30   tee: 1.00
15° taper: 0.60   wye, 30°: 0.18
30° taper: 1.00   45°: 0.28

coefficient of viscosity: VISCOSITY, ABSOLUTE
coefficient of viscosity of air: absolute viscosity of air   *
conservation of energy: Bernoulli's equation   *
Ⓧ **density:** mass per unit volume

English system, slug/ft^3

$$d = w/g = 0.0311w$$
$$= 4.476P/RT = 0.0839P/T$$

Metric system, gram/cm^3

AIR FLOW FORMULAS— *continued*

dynamic viscosity: VISCOSITY, ABSOLUTE
fan formulas: FAN FORMULAS
 **flow rate, laminar flow:**

by volume,

$$Q_A = 0.0069AV = 0.00545D^2V$$
$$= 0.0069Q_w RT/P = 0.368Q_w T/P$$
$$Q_S = 530Q_A/T$$

by weight,

$$Q_w = wQ_A = 0.0069wVA = 0.00545D^2wV$$
$$= 2.7Q_A P/T$$

flow rate, turbulent flow:

by volume,

$$Q_A = 141[(P_1^2 - P_2^2)D^5/fTL]^{1/2}$$

by weight,

$$Q_w = 10.56[(P_1^2 - P_2^2)D^5/fTL]^{1/2}$$

flow types, based on Reynold's number:

laminar, $N_R < 2000$
transitional, $2000 < N_R < 4000$
turbulent, $N_R > 4000$

friction factor (f): for round ducts, may be obtained from Moody diagram or other charts. Ducts other than round may be converted to equivalent round-duct diameter, where

$$D = 4A/s$$

friction factor (f):

approximate values for velocities between 1000 and 5000 ft/min in galvanized iron ducts,

D	4	6	8	10	12	18
f	0.023	0.020	0.019	0.018	0.017	0.0155

AIR FLOW FORMULAS — *continued*

D	24	36	48	60	84
f	0.0145	0.0135	0.013	0.012	0.009

friction loss: head loss, friction ∗
gas constant for air:

$$R = 53.35 \text{ ft-lb/lb/}°R$$

head, converted to pressure:

$$p = 0.036h$$

head, static pressure (h_S): static pressure energy head on the system based on pressure and weight of the air, or as measured with a manometer perpendicular to the flow.

$$h_S = h_T - h_V$$

head, total pressure (h_T):

$$h_T = h_S + h_V + h_L + h_F$$

(🔄) **head, velocity pressure (h_V):** kinetic energy, energy of the air due to its velocity,

$$h_V = h_T - h_S$$
$$= 8.3 \text{ E} - 07 \ V^2 w$$

head loss (h_L):

bends, $h_L = C_B h_V$
fittings, $h_L = C_F h_V$

(🔄) **head loss, friction (h_F):** based on duct length,

round duct,

$$h_F = 12 f h_V L/D$$
$$= 9.957 \text{ E} - 06 \ f L V^2 w/D$$

rectangular duct,

$$h_F = 3.98 \text{ E} - 05 \ f L V^2 sw/A$$

AIR FLOW FORMULAS — *continued*

kinematic viscosity: VISCOSITY, KINEMATIC
kinetic energy: head, velocity pressure   *
laminar flow: LAMINAR FLOW
Moody diagram: MOODY DIAGRAM
normal cubic meters per hour (NCMH): metric term used in the measurement of the flow of dry air at 0° Celsius and 1 bar atmospheric pressure.
perimeter (s):

rectangular duct,	$s = 2(a + b)$
round duct,	$s = 3.1416D$

pitot tube: PITOT TUBE
potential energy (h_P): energy due to the elevation of the system measured above datum.

pressure drop converted to head loss:

$$h_L = 27.76(p_1 - p_2)$$

pressure drop due to friction:

$$p_1 - p_2 = 3.6\,\text{E}-07\,fLV^2w/D$$

rectangular duct converted to round duct of equal friction:

$$D = 1.265(ab)^{3/5}/(a + b)^{1/5}$$

Reynold's number:

$$N_R = 2.07VDw/\mu$$
$$= 129VD/v$$

specific weight, actual:

$$w = 1.326P/T \;(P \text{ is in in. mercury})$$
$$= 0.0972P/T \;(P \text{ is in in. water})$$
$$= 2.7P/T \;(P \text{ is in psia})$$

specific weight of air at 70°F and atmospheric pressure:

$$w = 0.075 \text{ lb/ft}^3$$

AIR FLOW FORMULAS—*continued*

standard conditions:

$$\text{pressure} = 1 \text{ atm}$$
$$= 33.898 \text{ ft (water)}$$
$$= 29.92 \text{ in. (mercury)}$$
$$= 406.77 \text{ in. (water)}$$
$$= 14.696 \text{ psia}$$
$$\text{temperature} = 70°F$$
$$= 529.69°R$$

standard cubic feet per minute, scfm (Q_S): flow rate by volume at standard conditions, converted from actual flow rate (Q_A).

$$Q_S = 36 Q_A P/T$$

static head: head, static pressure   *
static pressure: pressure that tends to burst or collapse the pipe or duct.
static pressure head: head, static pressure   *
total pressure head: head, total pressure   *
transitional flow: TRANSITIONAL FLOW
turbulent flow: TURBULENT FLOW
velocity of flow:

$$V = 1098(h_V/w)^{1/2}$$
$$= 144 Q_A/A = 183.35 Q_A/D^2$$
$$= 144 Q_W/wA = 183.35 Q_W/D^2 w$$

velocity pressure: in the flow of the air, the pressure attributed to the velocity of the air.
velocity pressure head: head, velocity pressure   *
viscosity: VISCOSITY
volume change due to temperature: at constant pressure, the volume of a gas expands linearly with temperature change,

$$v_2 = v_1(T_2/T_1)$$

weight, specific: specific weight   *
weight, specific, of air: specific weight of air   *

air/fuel ratio for combustion of various fuels (ft^3 of air required):

acetylene: 177/lb

benzene: 177/lb

butane: 207/lb

carbon monoxide: 33/lb

coal: 170/lb

coke: 140/lb

ethane: 213/lb

ethyl alcohol: 120/lb

ethylene: 197/lb

fuel oil no. 2: 1400/gal

no. 3: 1420/gal

no. 5: 1460/gal

fuel oil no. 6: 1510/gal

gas, coke oven: 5/ft^3

manufactured: 4.3/ft^3

natural: 9.9′ft^3

hydrogen: 453/lb

methane: 230/lb

methyl alcohol: 85/lb

octane: 200/lb

pentane: 205/lb

propane: 208/lb

sulfur: 57/lb

toluene: 180/lb

AIR/GAS COMPRESSION FORMULAS: where

A = cylinder cross-sectional area, in.^{2}

c = clearance, decimal

C_P = specific heat at constant pressure, Btu/lb/°F

C_V = specific heat at constant volume, Btu/lb/°F

d = cylinder volume above piston top stroke, in.^{3}

D = piston displacement, in.^{3}/stroke

eff = volumetric efficiency, decimal

hp = horsepower

k = adiabatic exponent

ln = log to the base e

n = polytropic exponent

N = speed, rpm

P_1 = pressure, intake, psia

P_2 = pressure, discharge, psia

Q = compressor capacity, ft^3/min

r = compression ratio

s = number of stages

S = stroke, in.

T_1 = temperature, intake, °R

T_2 = temperature, exhaust, °R

V_1 = volume, intake, ft^3

V_2 = volume, discharge, ft^3

AIR/GAS COMPRESSION FORMULAS — *continued*

adiabatic compression: process without the addition or removal of heat,

$$P_1(V_1)^k = P_2(V_2)^k = \text{constant}$$

adiabatic exponent (k): the ratio of specific heat at contant pressure to the specific heat at constant volume,

$$k = C_P/C_V$$

adiabatic exponent (k) for various gases: THERMODYNAMICS

brake horsepower: horsepower at compressor input shaft.

capacity of compressor (Q): quantity compressed at intake pressure and temperature. In multistage compressors, capacity rating is the same as the first stage.

capacity reciprocating type compressor:

adiabatic process,

$$Q = ND(1 + c - c(r)^{1/k})/1728$$

clearance (c): in a reciprocating compressor, ratio of space (d) above the piston on the upstroke to piston displacement,

$$c = d/D$$

compression ratio (r): ratio of discharge pressure to intake pressure. For a multistage compressor, the same as the first stage.

$$r = P_2/P_1$$

displacement, piston, reciprocating compressor: volume displaced in the cylinder by the stroke of the piston from bottom to top.

$$D = AS$$

displacement, reciprocating compressor: same as the first-stage piston displacement.

efficiency, volumetric, reciprocating compressor: ratio of capacity of compressor to piston displacement.

$$\text{eff} = 1728Q/DN$$

efficiency, volumetric adiabatic:

$$\text{eff} = 1 + c - c(r)^{1/k}$$

AIR/GAS COMPRESSION FORMULAS — *continued*

horsepower (hp):

$$\text{adiabatic, hp} = 0.00436 s P_1 Q [r^{(k-1)/sk} - 1] k/(k-1)$$
$$\text{isothermal, hp} = 0.00436 P_1 Q (\ln r)$$
$$\text{polytropic, hp} = 0.00436 P_1 Q [r^{(n-1)/n} - 1] n/(n-1)$$

intercooling: removal of heat from air (or gas) between stages.

isentropic compression: adiabatic process, reversible, associated with reciprocating compressors.

isothermal compression: process where temperature remains constant,

$$P_1 V_1 = P_2 V_2 = \text{constant}$$

piston displacement: displacement, piston, reciprocating compressor   *

polytropic compression: process where friction heat is present and heat is removed.

$$P_1 (V_1)^n = P_2 (V_2)^n = \text{constant}$$

polytropic exponent (n):

$$n = 1.3 \text{ (approx. for air)}$$
$$= \log (P_1/P_2)/\log (V_2/V_1)$$

pressure, discharge (P_2):

$$\text{adiabatic, } P_2 = P_1 (V_1/V_2)^k = P_1 (T_2/T_1)^{k/(k-1)}$$
$$\text{isothermal, } P_2 = P_1 V_1/V_2$$
$$\text{polytropic, } P_2 = P_1 (V_1/V_2)^n = P_1 (T_2/T_1)^{n/(n-1)}$$

temperature, discharge (T_2):

$$\text{adiabatic, } T_2 = T_1 (r)^{(k-1)/k} = T_1 (V_1/V_2)^{k-1}$$
$$\text{isothermal, } T_2 = T_1$$
$$\text{polytropic, } T_2 = T_1 (r)^{(n-1)/n} = T_1 (V_1/V_2)^{n-1}$$

volume, discharge (V_2):

$$\text{adiabatic, } V_2 = V_1 (P_1/P_2)^{1/k} = V_1 (T_1/T_2)^{k/(1-k)}$$
$$\text{isothermal, } V_2 = P_1 V_1/P_2$$
$$\text{polytropic, } V_2 = V_1 (P_1/P_2)^{1/n} = V_1 (T_1/T_2)^{n/(1-n)}$$

volumetric efficiency: efficiency, volumetric

air horsepower: FAN FORMULAS
AISI: American Iron and Steel Institute
Al: aluminum (element)

Aluminum Wire Gauges (American Wire Gauge)

Size		Dia. (in.)	Electrical Resistance (ohms/1000 ft)	Weight (lb/1000 ft)
Solid	Stranded			
35		0.0056	543	0.0289
34		0.0063	429	0.0365
33		0.0071	338	0.0464
32		0.0080	266	0.0589
31		0.0089	215	0.0729
30		0.0100	170	0.0920
29		0.0113	133	0.118
28		0.0126	107	0.146
27		0.0142	84.5	0.186
26		0.0159	67.4	0.233
25		0.0179	53.2	0.295
24		0.0201	42.2	0.372
23		0.0226	33.4	0.470
22		0.0253	26.6	0.589
21		0.0285	21.0	0.748
20		0.0320	16.6	0.942
	20	0.0360	17.3	0.960
19		0.0359	13.3	1.19
18		0.0403	10.49	1.49
	18	0.046	10.9	1.53
17		0.0453	8.29	1.89
16		0.0508	6.60	2.38
	16	0.058	6.85	2.43
15		0.0571	5.23	3.00
14		0.0641	4.15	3.78
	14	0.073	4.31	3.87
13		0.0720	3.29	4.77

Aluminum Wire Gauges — *Continued*

Size		Dia. (in.)	Electrical Resistance (ohms/1000 ft)	Weight (lb/1000 ft)
Solid	Stranded			
12		0.0808	2.61	6.01
	12	0.092	2.71	6.16
11		0.0907	2.07	7.57
10		0.1019	1.64	9.556
	10	0.116	1.70	9.76
9		0.1144	1.30	12.04
	9	0.130	1.35	12.3
8		0.1285	1.03	15.2
	8	0.146	1.07	15.6
7		0.1443	0.818	19.16
	7	0.164	0.850	19.5
6		0.1620	0.649	24.15
	6	0.184	0.674	24.7
5		0.1819	0.515	30.45
	5	0.206	0.534	31.1
4		0.2043	0.408	38.41
	4	0.232	0.424	39.4
3		0.2294	0.324	48.43
	3	0.260	0.336	49.7
2		0.2576	0.257	61.07
	2	0.292	0.266	62.5
1		0.2893	0.203	77.0
	1	0.332	0.211	78.7
1/0		0.3249	0.161	97.1
	1/0	0.373	0.168	99.4
2/0		0.3648	0.128	122.0
	2/0	0.419	0.133	125.0
3/0		0.4096	0.101	154.0
	3/0	0.470	0.105	158.0
4/0		0.460	0.0805	195.0
	4/0	0.528	0.0836	199.0

Aluminum Wire Gauges — *Continued*

Size		Dia. (in.)	Electrical Resistance (ohms/1000 ft)	Weight (lb/1000 ft)
Solid	Stranded			
	250 mcm	0.575	0.0738	235.0
	300 mcm	0.630	0.0590	282.0
	350 mcm	0.681	0.0506	330.0
	400 mcm	0.728	0.0442	377.0
	450 mcm	0.772	0.0393	424.0
	500 mcm	0.813	0.0354	471.0
	550 mcm	0.855	0.0322	518.0
	600 mcm	0.893	0.0295	565.0
	650 mcm	0.929	0.0272	612.0
	700 mcm	0.964	0.0253	659.0
	750 mcm	0.998	0.0236	706.0
	800 mcm	1.031	0.0221	753.0
	900 mcm	1.094	0.0197	848.0
	950 mcm	1.125	0.0176	874.0
	1000 mcm	1.152	0.0177	942.0
	1250 mcm	1.289	0.0142	1177.0
	1500 mcm	1.412	0.0118	1412.0
	1750 mcm	1.526	0.0101	1648.0
	2000 mcm	1.632	0.00885	1883.0

Am: americium (element)

Amagat's Law: GAS LAWS

ambient temperature: temperature of the medium (gas or liquid) in its immediate surroundings.

American Wire Gauge (AWG): Brown and Sharpe wire gauge; gauge system for sizing bare aluminum, steel and copper wires.

ampere: ELECTRICAL

ampere =

0.1 abampere	1 E + 06 μA
1.0 C/s	1000 mA
1.036 E − 05 faraday/s	2.998 E + 09 statamperes
1.000157 International amperes	

ampere, International: quantity of current that deposits silver at the rate of 0.00118 g/s.

ampere, International = 0.999835 A, absolute

ampere/square centimeter = 6.4516 A/in.^{2}

ampere/square inch = 0.1550 A/cm^2

ampere-hour =

360 abcoulombs	0.03731 faraday
3600 A-s	1.079 E + 13 statcoulombs
3600 C	

ampere-second: COULOMB

ampere-turn: ELECTROMAGNETISM

ampere-turn =

0.1 abampere-turn
1.257 Gb

ampere-turn/centimeter =

2.54 A-turns/in.	1.257 Oe
1.257 Gb/cm	

ampere-turn/inch =

0.3937 A-turn/cm	0.313 line/in.^{2}
0.4947 Gb/cm	0.4947 Oe

ampere-turn/meter =

0.0254 A-turn/in.	0.01257 Oe
0.01257 Gb/cm	7.96E + 05 Wb/m^2

ANGLE, SECTION FORMULAS: where

a, b, c, d, x, y = dimensions, in.

A = area, in.^{2}

cg = center of gravity

I_{AA}, I_{BB}, I_{CC} = moment of inertia, in.^{4}

k_{AA}, k_{BB}, k_{CC} = radius of gyration, in.

t = thickness, in.

Z_{AA}, Z_{BB}, Z_{CC} = section modulus, in.^{3}

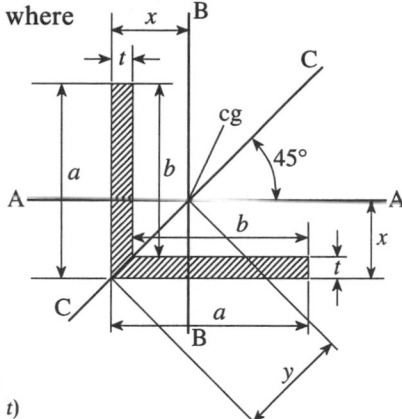

angle with equal legs:

area, $A = 2at - t^2$

center of gravity (cg) at

$$x = 0.5(a^2 + at - t^2)/(2a - t)$$
$$y = 0.707(a^2 + at - t^2)/(2a - t)$$

moment of inertia,

$$I_{AA} = I_{BB} = \frac{t(a - x)^3 + x^3a - b(x - t)^3}{3}$$
$$I_{CC} = 0.0833(t^3b + b^3t + 3a^2bt + t^4)$$

radius of gyration,

$$k_{AA} = (I_{AA}/A)^{1/2}$$
$$k_{BB} = (I_{BB}/A)^{1/2}$$
$$k_{CC} = (I_{CC}/A)^{1/2}$$

section modulus,

$$Z_{AA} = I_{AA}/(a - x) \quad \text{if } x < a/2$$
$$Z_{BB} = I_{BB}/(a - x) \quad \text{if } x < a/2$$
$$Z_{CC} = I_{CC}/y$$

angle with unequal legs:

area, $A = t(a + d)$

center of gravity (cg) at

$$x = 0.5(at + 2dt + d^2)/(a + d)$$
$$y = 0.5(bt + 2ct + c^2)/(b + c)$$

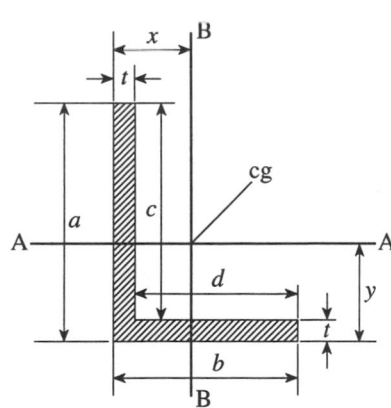

ANGLE, SECTION FORMULAS — *continued*

moment of inertia,

$$I_{AA} = \frac{t(a - y)^3 + y^3b - d(y - t)^3}{3}$$

$$I_{BB} = \frac{t(b - x)^3 + x^3a - c(x - t)^3}{3}$$

radius of gyration,

$$k_{AA} = (I_{AA}/A)^{1/2}$$
$$k_{BB} = (I_{BB}/A)^{1/2}$$

section modulus,

$$Z_{AA} = I_{AA}/(a - y) \quad \text{if } y < a/2$$
$$Z_{BB} = I_{BB}/(b - x) \quad \text{if } x < b/2$$

---

angle of repose: FRICTION

angstrom: small unit of length generally used in the measurement of wavelengths.

angstrom unit (Å) =

1 E − 08 cm	1 E − 07 mm
3.937 E − 09 in.	0.1 nm
1 E − 10 m	100 pm
1 E − 04 μm	

angular acceleration: ROTATION
angular displacement: ROTATION
angular impulse: ROTATION
angular momentum: ROTATION
angular velocity: ROTATION
annulus: CIRCLE SECTION FORMULAS; CIRCLE, HOLLOW
API: American Petroleum Institute
API degrees: DEGREES API
API gravity: DEGREES API

apothecary: weight system based on 8 drams per ounce and 12 ounces per pound; generally used by pharmacists.

apparent power: ELECTRICAL

arc cos, arc sin, etc.: TRIGONOMETRY

Archimedes' principle: BUOYANCY

are; ar: metric unit of measure for surface area

are: SQUARE DEKAMETER

arithmetic mean: average value of numbers, when the arithmetic mean is the sum of "n" numbers divided by "n".

arithmetic progression: addition of (*n*) number of terms, each term containing constant (*a*) plus (*nb*), as (*a* + *nb*), starting with $n = 0$ to $n = n - 1$, the final term, total

$$T = a + (a + b) + (a + 2b) + (a + 3b) + \cdots + [a + (n - 1)b]$$
$$T = 0.5n[2a + (n - 1)b]$$

As: arsenic (element)

ASA: American Standards Association

ASHVE: American Society of Heating and Ventilating Engineers

ASME: American Society of Mechanical Engineers

ASTM: American Society for Testing Materials

astronomical unit: mean distance from earth to the sun.

astronomical unit =

1.496 E + 08 km	1.496 E + 11 m
1.5803 E − 05 light-year	93 million mi

asymptote for a curve: when an equation of a curve is such that the numerical value of one unknown (*y*) increases without limit as the other unknown (*x*) approaches a value (*a*), then the line for where the value (*a*) is located is the asymptote for that curve.

Example, $x = 1$ for the curve $y = 2/(1 - x)$
then, $y = 2/0 = $ infinity and $a = 1$ is the asymptote

atmosphere =

1.01325 bar	2116.22 lb/ft^2
1.01325 E + 06 baryes	14.696 lb/in.^{2}
76 cm (mercury, 0°C, 32°F)	0.760 m (mercury, 0°C, 32°F)
1033.2 cm (water, 4°C, 39.2°F)	7.6 E + 05 μm (mercury, 0°C, 32°F)
1.01325 E + 06 dyn/cm^2	1013.25 mbar
33.898 ft (water, 4°C, 39.2°F)	1.01325 E – 06 Mbar
33.968 ft (water, 20°C, 68°F)	760 mm (mercury, 0°C, 32°F)
1033.23 g/cm^2	10,332 mm (water, 4°C, 39.2°F)
29.92 in. (mercury, 0°C, 32°F)	10,354 mm (water, 20°C, 68°F)
406.77 in. (water, 4°C, 39.2°F)	1.01325 E + 05 N/m^2
407.62 in. (water, 20°C, 68°F)	234.54 oz/in.^{2}
1.03323 kg/cm^2	1.01325 E + 05 Pa
10,332.3 kg/m^2	1.058 ton/ft^2
101.325 kPa	760 torr

atmospheric pressure: measured as pressure exerted by a 29.92-in. column of mercury at 0°C, 32°F temperature and equal to 14.696 lb/in.^{2}, normally referred to as atmosphere, barometric pressure, or standard atmospheric pressure.

atom: smallest particle of an element, with a mass of 1.66 E – 24g.

atomic number: refers to the atomic number of an element as per order of listing of elements in the Periodic Table.

atomic numbers of elements: ELEMENTS

atomic weight: unit weight system based on oxygen equal to atomic weight of 16 (units).

atomic weight of elements: ELEMENTS

atto: prefix equals 1 E – 18.

Au: gold (element)

autoignition temperature: temperature of a fuel or solvent at which it will self-ignite, as in a spontaneous reaction.

autotransformer: ELECTRICAL

Avogadro's gas law: GAS LAWS

Avogadro's number =

6.0234 E + 23 molecules/gram mole
2.732 E + 26 molecules/pound mole

avoirdupois: weight system based on 16 dr/oz and 16 oz/lb; used for drugs and precious metals.

AWG: American Wire Gauge

AWG aluminum wire sizes: ALUMINUM WIRE GAUGES

AWG copper wire sizes: COPPER WIRE GAUGES

AWG steel wire sizes: STEEL WIRE GAUGES

B

B: boron (element)
Ba: barium (element)
bar: metric unit for pressure
bar =

0.9869 atm
1 E+06 baryes
75.006 cm (mercurcy, 0°C, 32°F)
1 E+06 dyn/cm^2
33.45 ft (water, 4°C, 39.2°F)
1019.72 g/cm^2
29.53 in. (mercury, 0°C, 32°F)
401.45 in. (water, 4°C, 39.2°F)
402.29 in. (water, 20°C, 68°F)
1.0197 kg/cm^2
1.0197 E+04 kg/m^2
100 kPa

2088 lb/ft^2
14.5038 lb/in.^{2}
0.750 m (mercury, 0°C, 32°F)
1000 mbar
1 E−06 Mbar
1 E+06 μbar
750.06 mm (mercury, 0°C, 32°F)
1 E+05 N/m^2
1 E+05 Pa
1.044 ton/ft^2
750.06 torr

barn: unit for measurement of cross-sectional area in nuclear physics.
barn =

1 E−24 cm^2
1 E−28 m^2

barometer: instrument used for the measurement of atmospheric pressure.

barometric pressure: ATMOSPHERIC PRESSURE
barrel (dry) =

0.970 bbl (liquid)	7056 in.^{3}
3.281 bushels (level)	0.1156 m^3
4.083 ft^3	13.125 pecks
26.25 gal (dry)	105 qt (dry)

barrel (liquid) =

1.0313 bbl (dry)	31.5 gal (liquid)
4.211 ft^3	119.24 l
7276.5 in.^{3}	0.119 m^3
26.23 gal (Imperial)	

barrel (petroleum) =

5.6146 ft^3	159 l
42 gal	0.159 m^3

barye: DYNE/SQUARE CENTIMETER
Baumé scale: DEGREES BAUMÉ
BEAM LOAD FORMULAS: where

a, b, c, e, x = dimensions, in.
d = deflection at load, in.
d_C = deflection at center, in.
d_L = deflection at left end of beam, in.
d_R = deflection at right end of beam, in.
d_{max} = maximum deflection, in.
E = modulus of elasticity, lb/in.^{2}
I = moment of inertia, in.^{4}
L = distance between supports, in.
M = moment, in.-lb
M_C = moment at center, in.-lb
M_{max} = maximum moment, in.-lb
M_x = moment at point "x," in.-lb
P, P_1, P_2 = concentrated loads, lb
R_1, R_2 = reaction forces, lb
s = stress, maximum, lb/in.^{2}
  = M_{max}/Z

BEAM LOAD FORMULAS — *continued*

s_W = stress, working, lb/in.^{2}
  = s/sf
sf = safety factor, code or design
V_{max} = shear load, maximum, lb
V_x = shear load at point "x," lb
w = equal loading at w, lb/in.
W = total load, lb
Z = section modulus, in.^{3}
  = M_{max}/s

$d_{max} = P(L^3)/48EI$, at $x = 0.5L$
$d_x = Px(0.0625L^2 - 0.0833x^2)/EI$, if $x < 0.5L$
$M_{max} = 0.25PL$, at $x = 0.5L$
$M_x = 0.5Px$, if $x < 0.5L$
$M_x = 0.5P(x - 0.5L)$, if $x > 0.5L$
$R_1 = R_2 = 0.5P$
$V_{max} = 0.5P$

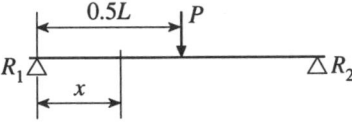

$d = 0.333P(ab)^2/EIL$, at $x_1 = a$
$d_{max} = 0.064Pa(Lb + ab)^{3/2}/EIL$
$M_{max} = Pab/L$, at $x_1 = a$
$M_{x1} = Pbx_1/L$, if $x_1 < a$
$M_{x2} = Pax_2/L$, if $x_2 < b$
$R_1 = Pb/L$
$R_2 = Pa/L$
$V_{max} = Pb/L$, if $a < b$
$V_{max} = Pa/L$, if $a > b$

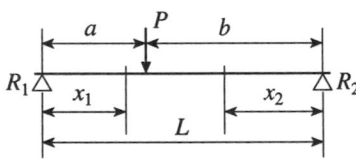

$d_L = 0.1667Pabc(L + b)/EIL$
$d_{max} = 0.064Pb(La + ab)^{3/2}/EIL$
$d_R = 0.1667Pabe(L + a)/EIL$
$M_{max} = Pab/L$, at $x = a$
$M_x = Pbx/L$, if $x < a$
$M_{x1} = Pb(L - x_1)/L$, if $a < x_1 < L$
$R_1 = Pb/L$
$R_2 = Pa/L$
$V_x = Pb/L$, if $x < a$
$V_{x1} = Pa/L$, if $a < x_1 < L$

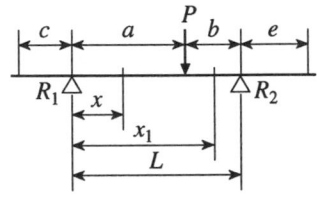

BEAM LOAD FORMULAS—*continued*

$d_{max} = 0.333a^2P(L + a)/EI$, at $x = a$
$d_{max} = 0.06415L^2aP/EI$, at $x_1 = 0.577L$
$M_{max} = Pa$, at R_1
$M_x = P(a - x)$, if $x < a$
$M_{x_1} = Pax_1/L$, if $x_1 < L$
$R_1 = P + Pa/L$
$R_2 = Pa/L$
$V_x = P$, if $x < a$
$V_{x1} = Pa/L$, if $x_1 < L$

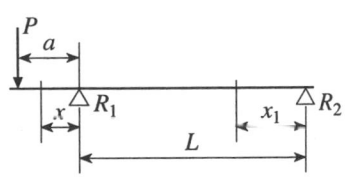

$d_L = 0.333a^2P(a + L)/EI$
$d_{max} = -0.0643L^2aP/EI$, at $x_1 = 0.4226L$
$d_R = 0.1667PabL/EI$
$M_x = P(a - x)$
$M_{x1} = Pa(L - x_1)/L$
$R_1 = P(a + L)/L$
$R_2 = -Pa/L$

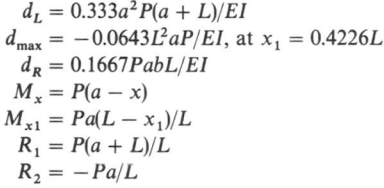

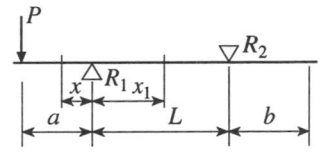

$d_{max} = 0.0417Pa(3L^2 - 4a^2)/EI$, at $x = 0.5L$
$d_x = 0.1667Px(3La - 3a^2 - x^2)/EI$, if $x < a$
$d_x = 0.1667a^2P(3L - 4a)/EI$, at $x = a$
$d_x = 0.1667Pa(3Lx - 3x^2 - a^2)/EI$, if $a < x < b$
$M_{max} = Pa$, if $a < x < b$
$M_x = Px$, if $x < a$
$M_x = Pa$, if $x > a$
$R_1 = R_2 = P$
$V_{max} = P$, if $x < a$

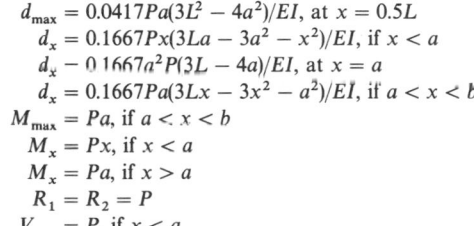

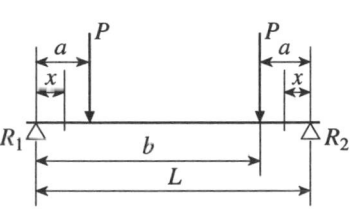

$M_{max} = Pa(L - a + b)/L$, if $a > b$
$M_{max} = Pb(L + a - b)/L$, if $b > a$
$M_x = Px(L - a + b)/L$, if $x < a$
$M_{x1} = R_1x_1 + Pa - Px_1$, if $a < x_1 < c$
$R_1 = P(L - a + b)/L$, if $a < b$
$R_2 = P(L + a - b)/L$, if $a > b$
$V_a = R_1$
$V_b = R_2$
$V_x = P(b - a)/L$, if $a < x < c$

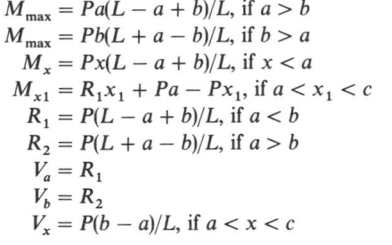

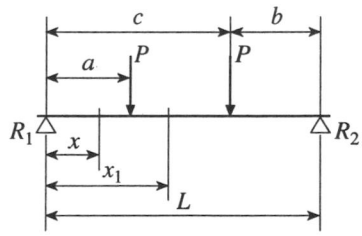

BEAM LOAD FORMULAS—*continued*

$M_{max} = R_1 a$, if $R_1 < P_1$
$M_{max} = R_2 b$, if $R_2 < P_2$
$M_x = (P_1 aL - P_1 ax + P_2 bx)/L$, if $a < x < c$
$R_1 = [P_1(L - a) + P_2 b]/L$
$R_2 = [P_1 a + P_2(L - b)]/L$
$V_a = R_1$
$V_b = R_2$
$V_x = R_1 - P_1$, if $a < x < c$

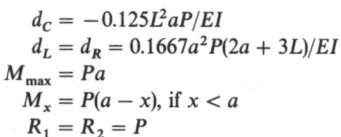

$d_C = -0.125 L^2 aP/EI$
$d_L = d_R = 0.1667 a^2 P(2a + 3L)/EI$
$M_{max} = Pa$
$M_x = P(a - x)$, if $x < a$
$R_1 = R_2 = P$

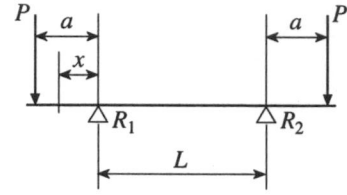

$M_{max} = P_1 a$ or $P_2 b$, evaluate each
$M_x = P_1(a - x)$, if $x < a$
$M_{x1} = P_1 a + (P_1 - R_1)x_1$, if $x_1 < L$
$M_{x2} = P_2(b - x_2)$, if $x_2 < b$
$R_1 = P_1 + (P_1 a - P_2 b)/L$
$R_2 = P_2 + (P_2 b - P_1 a)/L$
$V_x = P_1$
$V_{x1} = P_1 - R_1$
$V_{x2} = P_2$

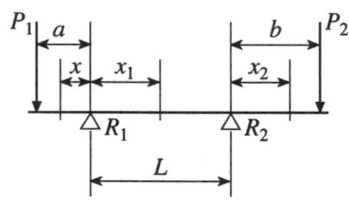

$M_x = R_1 x$, if $x < c$
$M_{x1} = R_1 x_1 - P_1(b + x_1 - L)$, if $c < x_1 < L$
$M_{x2} = P_2(a - x_2)$, if $x_2 < a$
$R_1 = (P_1 b - P_2 a)/L$
$R_2 = [P_1(L - b) + P_2(L + a)]/L$
$V_x = (bP_1 - aP_2)/L$
$V_{x1} = R_1 - P_1$
$V_{x2} = P_2$

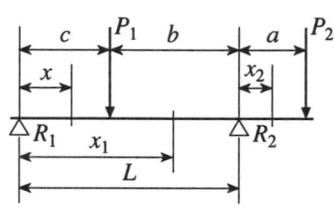

BEAM LOAD FORMULAS—*continued*

$d_{max} = 0.013L^4w/EI$, at $x = 0.5L$
$M_{max} = 0.125L^2w$, at $x = 0.5L$
$M_x = 0.5wx(L - x)$
$R_1 = R_2 = 0.5wL$
$V_{max} = 0.5wL$, at R_1, R_2
$V_x = w(0.5L - x)$

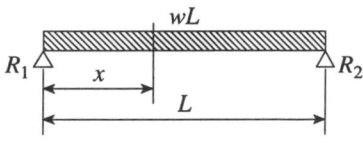

$d_C = [0.013L^4w - 0.0625(aL)^2w]/EI$
$d_L = d_R = wa(3a^3 + 6a^2L - L^3)/24EI$
$M_C = 0.125w(4a^2 - L^2)$
$M_x = 0.5w(a - x)^2$, if $x < a$
$M_{x1} = 0.5w(a^2 + x_1^2 - x_1L)$, if $x_1 < L$
$R_1 = R_2 = 0.5w(L + 2a)$

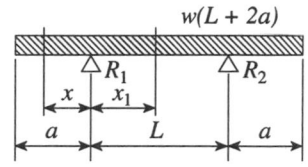

$d_L = wa(2b^2L + 4a^2L + 3a^3 - L^3)/24EI$
$d_R = wb(2a^2L + 4b^2L + 3b^3 - L^3)/24EI$
$M_{max} = R_1(R_1/2w) - R_1a$
$M_x = 0.5w(a - x)^2$, if $x < a$
$M_{x1} = 0.5w(a + x_1)^2 - R_1x_1$, if $x_1 < L$
$M_{x2} = 0.5w(b - x_2)^2$, if $x_2 < b$
$R_1 = \lceil w(a + L)^2 - b^2w\rceil/2L$
$R_2 = [w(b + L)^2 - a^2w]/2L$
$V_{max} = wa$

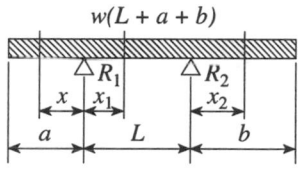

$M_{max} = 0.5R_1^2/w$
$M_x = 0.5wx(L^2 - a^2 - xL)/L$
$M_{x1} = 0.5w(a - x_1)^2$
$R_1 = 0.5w(L^2 - a^2)/L$
$R_2 = 0.5w(L + a)^2/L$
$V_x = R_1 - wx$, if $x < L$
$V_{x1} = w(a - x_1)$

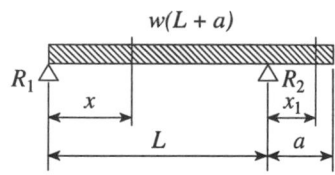

BEAM LOAD FORMULAS— *continued*

$M_{max} = 0.5R_1^2/w$
$M_x = R_1 x - 0.5x^2 w$, if $x < a$
$M_{x1} = R_2(L - x_1)$, if $x_1 > a$
$R_1 = 0.5wa(2L - a)/L$
$R_2 = 0.5a^2 w/L$
$V_{max} = R_1$
$V_x = R_1 - wx$, if $x < a$
$V_{x1} = R_2$, if $x_1 > a$

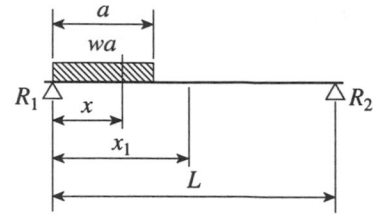

$M_{max} = R_1(a + 0.5R_1/w)$
$M_x = R_1 x$, if $x < a$
$M_x = R_1 x - 0.5w(x - a)^2$, if $a < x < e$
$M_x = R_2(L - x)$, if $x > e$
$R_1 = 0.5wb(2c + b)/L$
$R_2 = 0.5wb(2a + b)/L$
$V_a = R_1$
$V_c = R_2$
$V_x = R_1 + wa - wx$, if $a < x < e$

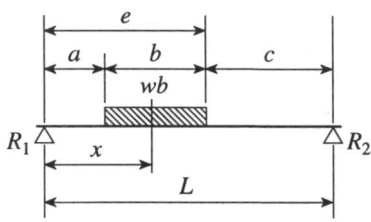

$d_{max} = 0.013044L^3 W/EI$, at $x = 0.519L$
$M_{max} = 0.1283WL$, at $x = 0.5774L$
$M_x = 0.333Wx(L^2 - x^2)/L^2$
$R_1 = 0.333W$
$R_2 = 0.667W$
$V_{max} = 0.667W$, at R_2
$V_x = W[0.333 - (x/L)^2]$

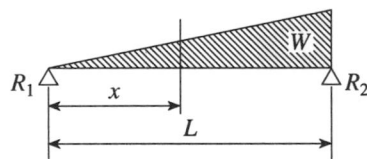

$d_{max} = 0.0167L^3 W/EI$, at $x = 0.5L$
$M_{max} = 0.1667WL$, at $x = 0.5L$
$M_x = Wx(0.5 - 0.667x^2)/L^2$, if $x < 0.5L$
$R_1 = R_2 = 0.5W$
$V_{max} = 0.5W$, at $x = R_1 = R_2$
$V_x = 0.5W(L^2 - 4x^2)/L^2$, if $x < 0.5L$

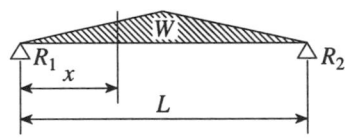

BEAM LOAD FORMULAS — *continued*

$d_{max} = 0.333L^3P/EI$, at P
$d_x = 0.1667Px^2(3L - x)/EI$
$M_{max} = PL$, at R_1
$M_x = Px$
$R_1 = P$
$V_{max} = P$

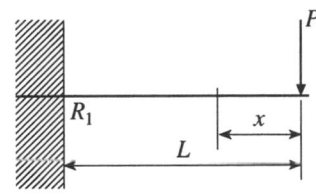

$d_x = 0.333a^3P/EI$, at $x = b$
$d_{max} = 0.1667a^2P(3L - a)/EI$, at $x = L$
$M_{max} = Pa$, at R_1
$M_x = P(x - b)$, if $x > b$
$R_1 = P$
$V_{max} = P$
$V_x = P$, if $x > b$

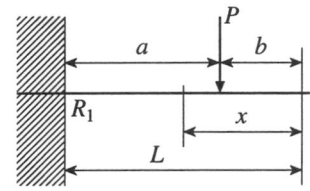

$d_x = 0.0091L^3P/EI$, at $x = 0.5L$
$d_{max} = 0.0093L^3P/EI$, at $x = 0.4472L$
$M_{max} = 0.1875PL$, at R_1
$M_x = 0.3125Px$, if $x < 0.5L$
$M_x = 0.15625PL$, at $x = 0.5L$
$M_x = 0.5PL - 0.6875Px$, if $x > 0.5L$
$R_1 = 0.6875P$
$R_2 = 0.3125P$
$V_{max} = 0.6875P$

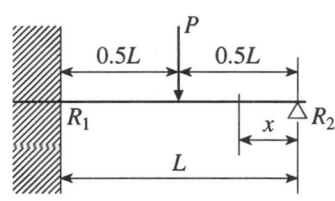

$d_{max} = 0.0098L^3P/EI$, at $b = 0.414L$
$M_{max} = R_2b$, at $x = b$
$M_x = 0.5Pab(b + L)/L^2$, at $x = L$
$M_x = R_2x$, if $x < b$
$M_x = R_2x - P(x - b)$, if $x > b$
$R_1 = 0.5Pb(3L^2 - b^2)/L^3$
$R_2 = 0.5a^2P(3L - a)/L^3$
$V_x = R_1$, if $x < b$
$V_x = R_2$, if $x > b$

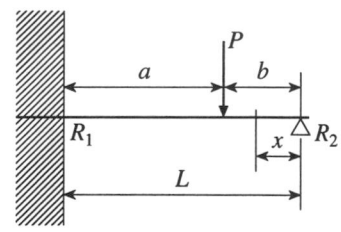

BEAM LOAD FORMULAS— *continued*

$d_{max} = 0.125L^4w/EI$
$M_{max} = 0.5L^2w$, at R_1
$M_x = 0.5x^2w$
$R_1 = wL$
$V_{max} = wL$, at R_1
$V_x = wx$

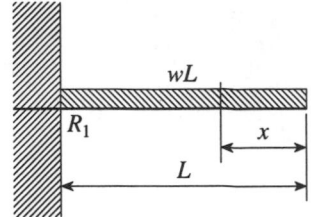

$d_C = 0.0052L^4w/EI$
$d_{max} = 0.0054L^4w/EI$, at $x = 0.4215L$
$M_{max} = 0.125L^2w$, at R_1
$M_x = 0.125wx(3L - 4x)$
$R_1 = 0.625wL$
$R_2 = 0.375wL$
$V_{max} = 0.625wL$, at R_1
$V_x = 0.375w(L - x)$

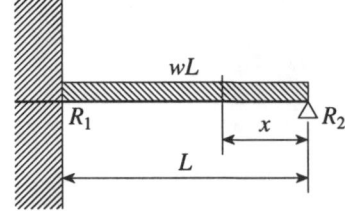

$d_{max} = 0.0667L^3W/EI$, at $x = 0$
$M_{max} = 0.333WL$, at R_1
$M_x = 0.333x^3W/L^2$
$R_1 = W$
$V_{max} = W$, at R_1
$V_x = x^2W/L^2$

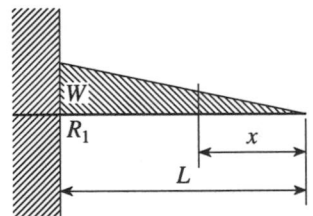

$d_{max} = 0.00477L^3W/EI$
$M_{max} = 0.1333WL$, at R_1
$M_x = Wx(0.2L^2 - 0.333x^2)L^2$
$R_1 = 0.8W$
$R_2 = 0.2W$
$V_{max} = 0.8W$, at R_1
$V_x = W(0.2L^2 - x^2)/L^2$

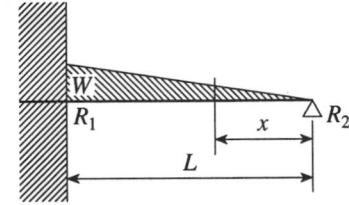

BEAM LOAD FORMULAS — *continued*

$d_{max} = 0.1833L^3W/EI$
$M_{max} = 0.6667WL$
$\quad M_x = 0.333x^2W(3L - x)/L^2$
$\quad R_1 = W$
$V_{max} = W$, at R_1
$\quad V_x = Wx(2L - x)/L^2$

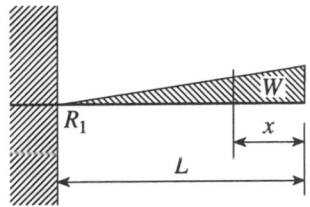

$d_{max} = 0.0052L^3P/EI$, at $x = 0.5L$
$\quad d_x = 0.0208x^2P(3L - 4x)/EI$
$M_{max} = 0.125PL$, at R_1, R_2, if $x = 0.5L$
$\quad M_x = 0.125P(4x - L)$, if $x < 0.5L$
$\quad M_x = 0.125P(4x - 3L)$, if $x > 0.5L$
$\quad R_1 = R_2 = 0.5P$
$V_{max} = 0.5P$

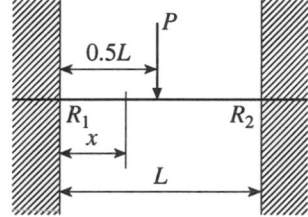

$d_{max} = 0.667a^2b^3P/EI(a + 3b)^2$, if $a < b$
$M_{max} = b^2aP/L^2$, at R_1 and if $a < b$
$\quad M_x = R_1x - b^2Pa/L^2$, if $x < a$
$\quad M_x = 2a^2b^2P/L^3$, at $x = a$
$\quad M_x = R_1x - P(x - a) - b^2Pa/L^2$, if $x > a$
$\quad R_1 = b^2P(3a + b)/L^3$
$\quad R_2 = a^2P(a + 3b)/L^3$
$\quad V_x = b^2P(3a + b)/L^3$, if $x < a$
$\quad V_x = a^2P(a + 3b)/L^3$, if $x > a$

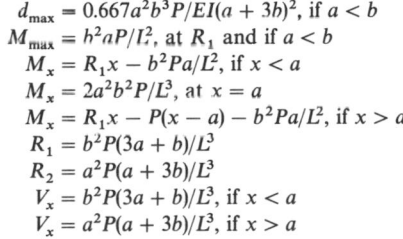

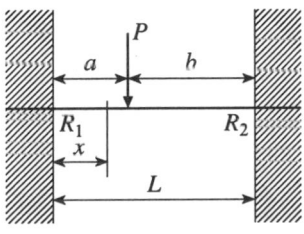

$d_{max} = 0.0026L^4w/EI$, at $x = 0.5L$
$\quad M_C = 0.0417L^2w$, at $x = 0.5L$
$M_{max} = 0.0833L^2w$, at R_1, R_2
$\quad M_x = 0.0833L^2w - 0.5Lwx + 0.5x^2w$
$\quad R_1 = R_2 = 0.5wL$
$V_{max} = 0.5wL$
$\quad V_x = 0.5wL - wx$

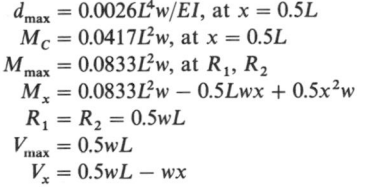

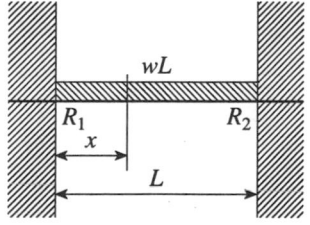

BEAM LOAD FORMULAS— *continued*

$d_{max} = 0.01183L^3W/EI$
$M_{max} = 0.043WL$, at $x = 0.548L$
  $M_x = 0.3Wx - 0.0667WL - 0.333x^3W/L^2$
  $R_1 = 0.3W$
  $R_2 = 0.7W$
$V_{max} = 0.7W$
  $V_x = 0.3W - x^2W/L^2$

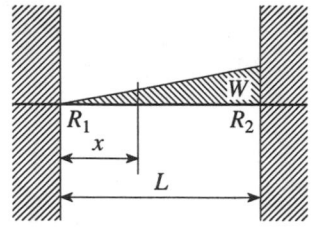

$d_{max} = 0.00156L^3W/EI$
$M_{max} = 0.0625WL$, at R_1, R_2
  $M_x = 0.5Wx - 0.0625WL - 0.333x^2W(3L - 2x)/L^2$
  $R_1 = R_2 = 0.5W$
$V_{max} = 0.5W$
  $V_x = 0.5W(L - 2x)^2/L^2$

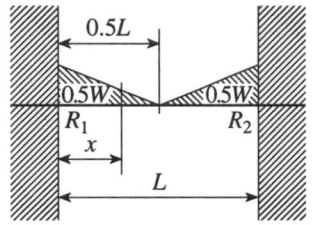

$d_{max} = 0.00625L^3W/EI$
$M_{max} = 0.0625WL$, at $x = 0.5L$
  $M_x = 0.1042WL + 0.5Wx - 0.667x^3W/L^2$
  $R_1 = R_2 = 0.5W$
$V_{max} = 0.5W$
  $V_x = 0.5W - 2x^2W/L^2$

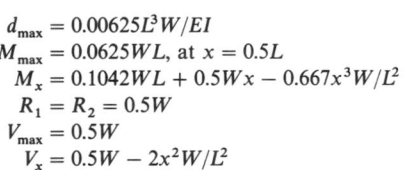
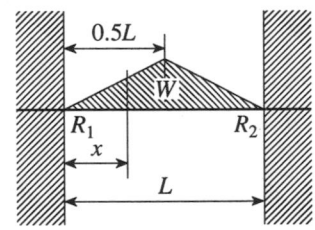

---

bel $= 10$ dB
belt length for pulley drives: PULLEY FORMULAS
bending stress: STRENGTH OF MATERIALS
benzine: petroleum ether
benzol: benzene

hydrofluoric acid: 248
hydrogen: −423
hydrogen bromide: −88.6
hydrogen chloride: −121
hydrogen fluoride: 68
hydrogen iodide: −33
hydrogen peroxide: 304
hydrogen sulfide: −76
indium: 3640
iodine: 364
iridium: 9570
iron: 5430
isobutane: 10.3
isopropyl alcohol: 176
kerosene: 510–575
krypton: −242
lanthanum: 6272
lead: 3180
linseed oil: 600
lithium: 2500
magnesium: 2030
magnesium chloride: 2574
magnesium oxide: 6580
manganese: 3750
mercury: 675
mercury chloride: 576
methane: −258
methanol: 149
methyl acetate: 138
methyl alcohol: 149
methyl chloride: −11
methyl ethyl ketone: 176
mineral oil: 680
mineral spirits: 302
molybdenum: 8690

Bernoulli's eq
BHN: Brinel
bhp: brake h
Bi: bismuth (
Bk: berkeliui
black body: r
Blasius form
block: RECTA
board-foot =

> 2360 cm^3
> 1 ft$^2 \times$ 1
> 0.083 ft^3

boiler horsep
boiling point
> is equal
boiling point

> acetic aci
> acetone:
> acetylene
> acrylic ac
> air: −31'
> alcohol:
> aluminun
> ammonia
> antimony
> argon: −
> arsenic: 1
> asphalt: '
> barium: ?
> benzene:
> benzoic a
> beryllium
> bismuth:

boiling point of various m

> cesium: 1238
> charcoal: 7600
> chlorine: −30
> chloroform: 142
> chromium: 4840
> cobalt: 5252
> columbium: 5970
> copper: 4660
> ester gum: 495
> ethane: −127
> ether: 95
> ethyl acetate: 167
> ethyl alcohol: 172
> ethyl bromide: 101
> ethyl chloride: 54
> ethyl iodide: 160
> ethylene: −155
> ethylene glycol: 387
> fluorine: −305
> formic acid: 213
> freon F12: −21
> freon F22: −41.4
> fuel oil no. 1: 510–575
>   no. 2: 600–650
>   no. 3: 700–775
> gallium: 3990
> gasoline: 158–194
> germanium: 5142
> glycerine: 554
> gold: 5370
> graphite: 7600
> hafnium: >6700
> helium: −452
> heptane: 209
> hexane: 157

boiling point of various materials (°F) — *continued*

naphtha: 310
naphthalene: 425
naphtha VMP: 203
neon: −411
nickel: 4950
nitric acid: 186
nitric oxide: −241
nitrogen: −320
nitrous oxide: −129
octane: 256
oil: 662
oleic acid: 547
olive oil: 570
osmium: >9570
oxygen: −297
ozone: −170
palladium: 5320
palmitic acid: 532
paraffin: 662–806
pentane: 97
phenol: 360
phosgene: 46
phosphorus, white: 556
pitch: 325
platinum: 7770
potassium: 1400
propane: −44
propionic acid: 237
propylene: −54
radium: 2084
radon: −79
rubidium: 1292
ruthenium: >4890
selenium: 1274

silicon: 4712
silver: 3942
sodium: 1616
sodium chloride: 2545
sodium nitrate: 716
stearic acid: 556
strontium: 2500
styrene: 293
sulfur: 832
sulfur dioxide: 14
sulfuric acid: 640
tantalum: >7412
tellurium: 2534
thallium: 3000
thorium: >5432
tin: 4118
titanium: >5432
toluene: 231
toluol: 231
trichloroethylene: 189
tungsten: 10526
turpentine: 320
uranium: 7100
vanadium: 5932
vinyl chloride: 7
water, distilled: 212
wood alcohol: 149
xenon: −164
xylene: 287
xylol: 275
zinc: 1664
zinc chloride: 1350
zirconium: 9122

bolt of cloth =

3657.6 cm long   36.576 m long
120 ft long   40 yd long
1440 in. long

Boyle's law: GAS LAWS
Br: bromine (element)
brake horsepower (bhp): actual mechanical energy supplied.

bhp = horsepower output/efficiency

brightness: ILLUMINATION
Brinell Hardness Number (BHN): a reference number for the compara-
tive hardness of materials as determined by direct-readout testing
equipment or calculations. The number is based on the diameter of
the impression on the material made by a 10-mm ball pressed into
the material.
British thermal unit (Btu): unit for heat energy; the amount of heat
required to raise the temperature of 1 lb of pure water 1° Fahrenheit and is equivalent to 778.26 (ft-lb) of mechanical energy.
Brown and Sharpe wire gauge: AMERICAN WIRE GAUGE
Btu =

1.0406 E + 04 cm³-atm   9336 in.-lb
1.0544 E + 10 erg   1055 J
0.36747 ft³-atm   0.252 kg-cal
778.26 ft-lb   107.5 kg-m
2.502 E + 04 ft-pdl   1.054 kJ
252 g-cal   2.929 E − 04 kWh
1.075 E + 07 g-cm   10.405 l-atm
3.928 E − 04 hp-h   0.2929 Wh
3.984 E − 04 hp-h (metric)   1054.4 W-s

Btu/cubic foot = 3.728 E + 04 J/m³
Btu/cubic foot/°F = 6.707 E + 04 J/m³/K

Btu/gallon $= 2.787 \text{ E} + 05 \text{ J/m}^3$

Btu/hour: equivalent to 0.006972 lb of ice melted per hour or 8.33 E−05 ton of refrigeration.

Btu/hour =

0.01667 Btu/min	2.986 E−05 hp (boiler)
2.778 E−04 Btu/s	3.98 E−04 hp (metric)
2.929 E+06 erg/s	0.2929 J/s
778.26 ft-lb/h	0.252 kg-cal/h
12.96 ft-lb/min	1.79 kg-m/min
0.2162 ft-lb/s	2.929 E−04 kW
0.070 g-cal/s	0.2929 W
3.928 E−04 hp	

Btu/hour/square foot =

2.71 kg-cal/h/m²
3.15248 W/m²

Btu/hour-square foot-°F =

1.356 E−04 g-cal/s-cm²-°C
4.883 kg-cal/h-m²-°C

Btu/minute

60 Btu/h	0.0018 hp (boiler)
0.01667 Btu/s	0.02389 hp (metric)
1.75725 E+08 erg/s	17.5725 J/s
4.668 E+04 ft-lb/h	15.12 kg-cal/h
778.26 ft-lb/min	107.54 kg-m/min
12.97 ft-lb/s	0.01757 kW
4.2 g-cal/s	0.005 ton of refrigeration
0.02356 hp	17.570 W

Btu/minute/square foot $= 189.1489 \text{ W/m}^2$

Btu/pound =

22.9405 cm³-atm/g	3.9275 E − 04 hp-h/lb
0.36747 ft³-atm/lb	2.324 J/g
778.26 ft-lb/lb	2324 J/kg
0.5556 g-cal/g	2.324 kJ/kg

Btu/pound/°F =

1 g-cal/g/°C	4186.7 J/kg/K

Btu/pound/°R = 1 g-cal/g/K
Btu/second =

3600 Btu/h	1.4334 hp (metric)
60 Btu/min	1054 J/s
1.4335 Cheval-vapeur	907 kg-cal/h
1.054 E + 10 erg/s	15.12 kg-cal/min
2.8 E + 06 ft-lb/h	0.252 kg-cal/s
4.668 E + 04 ft-lb/min	6452 kg-m/min
778.26 ft-lb/s	107.57 kg-m/s
252 g-cal/s	1.0544 kW
1.415 hp	0.30 ton of refrigeration
0.108 hp (boiler)	1054.4 W

Btu/second/square foot = 1.1349 E + 04 W/m²
Btu/second-square foot-°F = 2.0428 E + 04 W/m²/K
Btu/second/square inch = 1.634 E + 06 W/m²
Btu/square foot =

0.27125 g-cal/cm²	1.1349 E + 04 J/m²

Btu/square foot/°F = 2.0428 E + 04 W/m²/K
Btu/square foot-hour-°F = 4.88 kg-cal/h-m²-°C
Btu content of various materials: HEAT VALUE OF VARIOUS MATERIALS
Btu-foot/hour-square foot-°F =

12 Btu-in./h-ft²-°F	1.7296 W-m/m²/K

Btu-inch/hour-square foot-°F =

0.0833 Btu-ft/hr-ft²-°F
3.445 E − 04 g-cal-cm/s-cm²-°C
0.00144 J-cm/s-cm²-°C
0.293 J-in./s-ft²-°F
12.4 kg-cal-cm/h-m²-°C
0.124 kg-cal-m/h-m²-°C

0.0144 kWh-cm/h-m²-°C
2.929 E − 04 kWh-in./h-ft²-°F
1.44 E − 04 kWh-m/h-m²-K
0.144 Wh-m/h-m²-K
0.00144 W-s-cm/s-cm²-°C

Btu (mean): 1/180 of the amount of heat required to raise the temperature of 1 lb of water from 32°F to 212°F.

bulk modulus: the ratio of the increase of the hydrostatic pressure to the decrease in the volume of a material; it is the reciprocal of its compressibility factor k, where

B = bulk modulus, lb/in.²/in.³
k = compressibility factor, volume change per psi
$P_2 - P_1$ = increase in hydrostatic pressure, psi
V_1 = initial volume, in.³
V_2 = final volume, in.³

bulk modulus:

$$B = \frac{(P_2 - P_1)V_1}{V_2 - V_1}$$

hydrostatic pressure change:

$$P_2 - P_1 = B(V_2 - V_1)/V_1$$

bulk modulus (B) of various materials (1 E + 06 lb/in.²):

aluminum: 10
brass: 8.5
copper: 17
iron, cast: 14
     wrought: 21

lead: 1.1
oil: 0.25
steel: 23
water: 0.3

Bunker B oil: fuel oil no. 5

Bunker C oil: fuel oil no. 6

buoyancy: per Archimedes' principle, an object immersed in a fluid is buoyed up with a force equal to the weight of the displaced fluid, where

F = buoyant force, lb
V = volume of fluid displaced, ft^3
w = specific weight of fluid, lb/ft^3

buoyant force of floating object:

$$F = Vw$$

bushel (heaped) = 1.27 bushels (level)

bushel (level) =

0.3048 bbl (dry)	35.239 l
0.787 bushel (heaped)	0.03524 m^3
35,239 cm^3	1192 oz (liquid)
35.24 dm^3	4 pecks
1.2445 ft^3	64 pt (dry)
8 gal (dry)	74.473 pt (liquid)
7.8146 gal (Imperial)	32 qt (dry)
9.309 gal (liquid)	37.24 qt (liquid)
2150 in.^{3}	0.035238 stere
0.03524 kl	0.04609 yd^3

BWG: Birmingham Wire Gauge

C

C: carbon (element)
Ca: calcium (element)
cable length =

120 fathoms	0.1185 mi (naut)
720 ft	0.1364 mi (stat)
219.456 m	

cable sag: CATENARY FORMULAS
caliber: unit of measure used for sizing diameter of ammunition and equals 0.01 in.
calorie: GRAM-CALORIE
calorie/mole: GRAM-CALORIE/MOLE
calorie-gram: GRAM-CALORIE
calorie-kilogram: KILOGRAM-CALORIE
candela: ILLUMINATION
candle: ILLUMINATION
candle/square centimeter =

6.4516 cd/in.^{2}	3.1416 L
1 E+04 cd/m^2	3141.6 mL
2914 ft-L	

candle/square foot =

0.00694 cd/in.^{2}	3.1416 ft-L
10.7639 cd/m^2	0.0033816 L

candle/square inch =

0.155 cd/cm^2	452 ft-L
144 cd/ft^2	0.48695 L
1550 cd/m^2	486.95 mL

candle/square meter =

0.0929 cd/ft^2	0.2919 ft-L
6.4516 E−04 cd/in.^{2}	3.1416 E−04 L

candlepower: ILLUMINATION
candlepower (spherical) = 12.566 lm
capacitance: ELECTRICAL
capactive reactance: ELECTRICAL
capacitivity: ELECTROSTATICS
capacitor: ELECTRICAL
capillary action: the ability of a liquid (in a small-diameter tube) to rise in the tube due to surface tension of the liquid and the strength of the film adhesion between the liquid and the tube.
carat (gold): measure of parts of gold per 24 parts of an alloy and equal to 41.667 milligrams of gold per gram of alloy.
carat (precious stones) =

0.2 g	2 E−04 kg
3.08647 gr	200 mg

catenary: the sagging curve formed by a cable of uniform weight, suspended freely between two supports at the same level.

CATENARY FORMULAS: where

d = outside diameter of cable, in.
d_i = outside diameter of ice-covered cable, in.
D = distance between supports, ft

CATENARY FORMULAS — *continued*

h = sag height of cable, ft
L = length of cable, ft
p = wind pressure projected on cable, lb/ft^2
t = thickness of ice, in.
T = tension (horizontal), lb
w = weight of cable, lb/ft
w_i = weight of ice on cable, lb/ft
w_w = weight results due to wind pressure, lb/ft
W = total weight of cable, lb/ft

cable sag: sag height of cable   *
diameter of cable with ice:
$$d_1 = d + 2t$$

length of cable:
$$L = D + 2.667h^2/D$$

sag height of cable:
$$h = 0.125D^2W/T$$

tension in cable:
$$T = 0.125D^2W/h$$

weight, due to wind pressure:

no ice on cable, $w_w = 0.0833pd$
with ice on cable, $w_w = 0.0833pd_i$

weight of cable, total:
$$W = w + w_i + w_w$$

weight of ice on cable:
$$w_i = 0.306(d_i^2 - d^2)$$

wind pressure (p) on cable at various velocities:

wind velocity, mi/hr:	40	50	60	70	75
wind pressure (p), psf:	4	6	9	12	14

Cb: columbium (element)
Cd: cadmium (element)
Ce: cerium (element)
Celsius: (formerly Centigrade) temperature scale with water's freezing point set at zero and water's boiling point set at 100.
Celsius degrees: DEGREES C
cental =

   45.359 kg
   100 lb (force)

centare, centiare: SQUARE METER
center of gravity (cg); center of mass: point on a body whereby all forces pass through and the algebraic sum of all moments of areas about the axis through that point equals zero.
center of percussion: ROTATION
centi (c): prefix, equals 1 E−02
centiare: CENTARE
Centigrade: temperature term replaced by Celsius.
centigram =

   0.1 dg   1 E−05 kg
   0.01 g   10 mg
   0.1543 gr

centiliter =

   10 cm³   0.01 l
   0.1 dl   10 ml
   0.610 in.³   0.338 oz (liquid)

centimeter =

   1 E+08 Å   0.001 dam
   4.971 E−04 chain (Gunter)   0.1 dm
   3.281 E−04 chain (Ramden)   0.00875 ell
   0.02187 cubit   0.0055 fathom

centimeter — *continued*

0.03281 ft	6.214 E − 06 mi (statute)
0.0984 hand	393.7 mil
1 E − 04 hm	10 mm
0.3937 in.	1 E + 07 nm
1 E − 05 km	2.371 picas (printer's)
0.04971 link (Gunter)	1 E + 10 pm
0.03281 link (Ramden)	28.453 points (printer's)
0.01 m	0.00199 rod
1 E + 04 μm	0.01094 yd
5.4 E − 06 mi (nautical)	

centimeter (mercury, 0°C, 32°F) =

0.01316 atm	136 kg/m^2
0.01333 bar	1.3332 kPa
1.3332 E + 04 dyne/cm^2	27.845 lb/ft^2
0.033281 ft (mercury)	0.19337 lb/in.^{2}
0.4460 ft (water, 4°C, 39.2°F)	10 mm (mercury)
13.5951 g/cm^2	1333.2 N/m^2
0.3937 in. (mercury)	1333.2 Pa
5.3522 in. (water, 4°C, 39.2°F)	10 torr
0.013595 kg/cm^2	

centimeter (water, 4°C, 39.2°F) =

9.679 E − 04 atm	0.014223 lb/in.^{2}
9.80648 E − 04 bar	0.7355 mm (mercury, 0°C, 32°F)
0.07355 cm (mercury, 0°C, 32°F)	98.0638 Pa
980.64 dyn/cm^2	0.7355 torr

centimeter/second =

1.969 ft/min	0.01944 knot (speed)
0.03281 ft/s	0.6 m/min
0.036 km/h	0.02237 mi/h
0.0006 km/min	3.728 E − 04 mi/min

centimeter/second/second =

0.03281 ft/s²	0.01 m/s²
0.036 km/h/s	0.02237 mi/h/s

centimeter-dyne; dyne-centimeter: ERG
centimeter-gram: GRAM-CENTIMETER
centimeter-second/gram: RHE
centimeter square/second: STOKE
centipoise (cP): metric unit of absolute (dynamic) viscosity and is equal to centistokes multiplied by density in g/cm³.
centipoise =

0.01 dyn-s/cm²	1 E+04 μP
0.01 g/cm/s	1 mPa-s
0.001 kg/m/s	0.001 N-s/m²
1.0197 E−04 kg-s/m²	0.01 P
2.088 E−05 lb (force)-s/ft²	0.001 Pa-s
1.45 E−07 lb (force)-s/in.²	6.7197 E−04 pdl-s/ft²
2.42 lb (mass)/ft/h	0.0752 slug/ft/h
0.04032 lb (mass)/ft/min	2.088 E−05 slug/ft/s
6.7197 E−04 lb (mass)/ft/s	

centistoke: cSt; metric unit of kinematic viscosity and is equal to centipoise divided by density in g/cm³.
centistoke =

0.01 cm²/s	1 mm²/s
0.03875 ft²/h	2.16 SSF
1.0764 E−05 ft²/s	0.226 SSU − 195/SSU, if SSU < 100 s
0.00155 in.²/s	0.220 SSU − 135/SSU, if SSU > 100 s
1 E−06 m²/s	0.01 St

centrifugal force: ROTATION
centripetal force: ROTATION
centroid: same as center of gravity; generally used in reference to a line, surface, or volume. Centroids for curves and irregular surface areas are located with the use of calculus.

Cf: californium (element)
cgs: centimeter-gram-second
chain (Engineer): CHAIN (RAMDEN)
chain (Gunter) =

0.66 chain (Ramden)	100 links (Gunter)
1 chain (surveyor)	66 links (Ramden)
2011.68 cm	20.12 m
66 ft	0.0125 mi (statute)
0.1 furlong	4 rods
792 in.	22 yd
0.02012 km	

chain (Ramden) =

1 chain (engineer)	151.5 links (Gunter)
1.515 chain (Gunter)	100 links (Ramden)
3048 cm	30.48 m
100 ft	0.01894 mi (statute)
0.1515 furlong	6.06 rods
1200 in.	33.33 yd

chain (surveyor): CHAIN (GUNTER)
CHANNEL, SECTION FORMULAS: where

$a, b, c, d, e,$
m, n, t, x, y = dimensions, in.
A = area, in.^{2}
cg = center of gravity
I_{AA}, I_{BB} = moment of inertia, in.^{4}
k_{AA}, k_{BB} = radius of gyration, in.

formed channel:

area, $A = bd - ac$
center of gravity (cg),

$$x = \frac{2b^2m + t^2c}{2(bd - ac)}$$

$$y = 0.5d$$

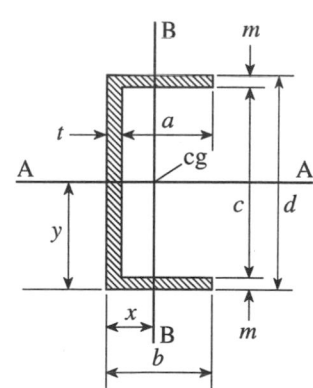

CHANNEL, SECTION FORMULAS — *continued*

moment of inertia,

$$I_{AA} = 0.0833(d^3b - c^3a)$$

$$I_{BB} = \frac{x^3d - d(x - t)^3 + 2m(b - x)^3}{3}$$

radius of gryration,

$$k_{AA} = \frac{0.2887(d^3b - c^3a)^{1/2}}{(bd - ac)^{1/2}}$$

$$k_{BB} = \frac{0.5774[x^3d - d(x - t)^3 + 2m(b - x)^3]^{1/2}}{(bd - ac)^{1/2}}$$

section modulus,

$$Z_{AA} = 0.1667(d^3b - c^3a)/d$$
$$Z_{BB} = I_{BB}/(b - x), \text{ if } x < 0.5b$$

standard channel:

area, $A = dt + a(m + n)$
center of gravity (cg),

$$x = \frac{0.5(t^2d) + 2am(t + 0.5a) + a(m - n)(t + 0.33a)}{dt + a(m + n)}$$

$$y = 0.5d$$

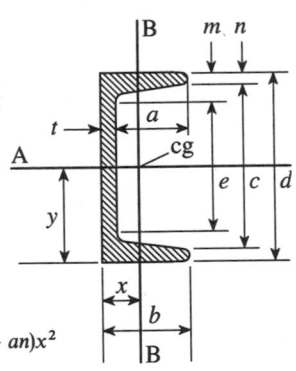

moment of inertia,

$$I_{AA} = 0.0833(d^3b) - \frac{0.01042a(c^4 - e^4)}{m - n}$$

$$I_{BB} = \frac{2b^3n + t^3e}{3} + \frac{(m - n)(b^4 - t^4)}{6a} - (dt + am + an)x^2$$

radius of gyration,

$$k_{AA} = (I_{AA}/A)^{1/2}$$
$$k_{BB} = (I_{BB}/A)^{1/2}$$

CHANNEL, SECTION FORMULAS — *continued*

section modulus,

$$Z_{AA} = I_{AA}/0.5d$$
$$Z_{BB} = I_{BB}(b - x), \text{ if } x < 0.5b$$

Charles' law: GAS LAWS
chemical formulas for various compounds:

acetaldehyde: CH_3CHO
acetic acid: $C_2H_4O_2$
acetone: C_3H_6O
acetylene: C_2H_2
aluminum oxide: Al_2O_3
ammonia: NH_3
benzene: C_6H_6
boric acid: H_3BO_3
butane: C_4H_{10}
calcium carbonate: $CaCO_3$
calcium chloride: $CaCl_2$
calcium fluoride: CaF_2
camphor: $C_{10}H_{16}O$
carbolic acid: C_6H_5OH
carbon dioxide: CO_2
carbon disulfide: CS_2
carbonic acid: H_2CO_3
carbon monoxide: CO
carbon tetachloride: CCl_4
chloroform: $CHCl_3$
ethane: C_2H_6
ether: $C_4H_{10}O$
ethyl alcohol: C_2H_5OH
ethyl chloride: C_2H_5Cl
ethylene: C_2H_4
ethylene glycol: $C_2H_6O_2$
ethyl ether: $C_4H_{10}O$

formaldehyde: $HCHO$
formic acid: CH_2O_2
freon F12: CCl_2F_2
freon F22: $CHClF_2$
glycerine: $C_3H_8O_3$
hematite: Fe_2O_3
heptane: C_7H_{16}
hexane: C_6H_{14}
hydrochloric acid: HCl
hydrocyanic acid: HCN
hydrofluoric acid: HF
hydrogen chloride: HCl
hydrogen fluoride: HF
hydrogen peroxide: H_2O_2
hydrogen sulfide: H_2S
isobutane: $(CH_3)_3CH$
isopropyl alcohol: $(CH_3)_2CHOH$
lead sulfide: PbS
methane: CH_4
methyl alcohol: CH_3OH
methyl chloride: CH_3Cl
naphthalene: $C_{10}H_8$
nitric acid: HNO_3
nitroglycerine: $C_3H_5(ONO_2)_3$
octane: C_8H_{18}
pentane: C_5H_{12}
phenol: C_6H_6O

chemical formulas for various compounds — *continued*

phosphoric acid: H_3PO_4

propane: C_3H_8

pyrite: FeS_2

silica acid: H_4SiO_4

sodium carbonate: Na_2CO_3

sodium chloride: NaCl

sodium hydroxide: NaOH

sodium nitrate: $NaNO_3$

sodium sulfate: Na_2SO_4

sulfur dioxide: SO_2

sulfuric acid: H_2SO_4

sulfurous acid: H_2SO_3

tartaric acid: $C_4H_6O_6$

toluene: C_7H_8

trichloroethylene: C_2HCl_3

trisodium phosphate: Na_3PO_4

turpentine: $C_{10}H_{16}$

water: H_2O

xylene: C_8H_{10}

Cheval-vapeur: HORSEPOWER (METRIC)

circle =

360°

400 grades

21,600 min

6.2832 rad

CIRCLE, SECTION FORMULAS: where

a = included angle, degrees

A = area, in.^{2}

b, y = dimensions, in.

c = chord length, in.

C = circumference, in.

cg = center of gravity

d = inside diameter, in.

D = outside diameter, in.

h = height of segment. in.

I_{AA}, I_{BB}, I_{CC} = moment of inertia, in.^{4}

I_P = polar moment of inertia, in.^{4}

k_{AA}, k_{BB}, k_{CC} = radius of gyration, in.

k_P = polar radius of gyration, in.

L = length of arc. in.

r = inside radius, in.

R = outside radius. in.

Z_{AA}, Z_{BB}, Z_{CC} = section modulus, in.^{3}

Z_P = polar section of modulus for torsion, in.^{3}

annulus: circle, hollow   ∗

CIRCLE, SECTION FORMULAS — *continued*

circle:

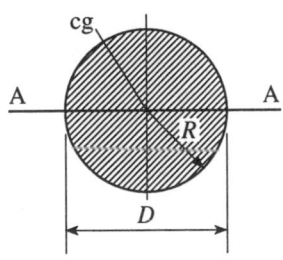

area,

$$A = 0.7854D^2 = 3.1416R^2$$

center of gravity (cg), at center of circle circumference,

$$C = 3.1416D = 6.2832R$$

diameter, $D = 2R = 0.31831C$
moment of inertia,

$$I_{AA} = 0.0491D^4$$
$$= 0.7854R^4$$

moment of inertia, polar,

$$I_P = 0.09818D^4 = 1.5708R^4$$

radius, $R = 0.5D = 0.159155C$
radius of gyration,

$$k_{AA} - 0.25D = 0.5R$$

radius of gyration, polar,

$$k_P = 0.3536D = 0.7071R$$

section modulus,

$$Z_{AA} = 0.09818D^3$$
$$= 0.7854R^3$$

section modulus, polar,

$$Z_P = 0.19635D^3$$
$$= 1.5708R^3$$

CIRCLE, SECTION FORMULAS — *continued*

circle (semi):

arc length,

$$L = 1.5708D = 3.1416R$$

area,

$$A = 0.3927D^2 = 1.5708R^2$$

center of gravity (cg),

$$y = 0.212D = 0.424R$$

diameter, $D = 2R$
moment of inertia,

$$I_{AA} = 0.00686D^4 = 0.1098R^4$$
$$I_{BB} = 0.0245D^4 = 0.392R^4$$
$$I_{CC} = 0.0245D^4 = 0.392R^4$$

moment of inertia, polar,

$$I_P = 0.0314D^4 = 0.5025R^4$$

radius, $R = 0.5D$
radius of gyration,

$$k_{AA} = 0.1322D = 0.2643R$$
$$k_{BB} = 0.25D = 0.50R$$
$$k_{CC} = 0.25D = 0.50R$$

radius of gyration, polar,

$$k_P = 0.283D = 0.566R$$

section modulus,

$$Z_{AA} = 0.0238D^3 = 0.19069R^3$$
$$Z_{BB} = 0.049D^3 = 0.392R^3$$
$$Z_{CC} = 0.049D^3 = 0.392R^3$$

CIRCLE, SECTION FORMULAS — *continued*

circle, hollow (annulus):

area,

$$A = 0.7854(D^2 - d^2)$$
$$= 3.1416(R^2 - r^2)$$

center of gravity (cg), at center of circle
diameter,

$$D = 2R$$
$$d = 2r$$

moment of inertia,

$$I_{AA} = 0.0491(D^4 - d^4)$$
$$= 0.7854(R^4 - r^4)$$

moment of inertia, polar,

$$I_P = 0.09818(D^4 - d^4)$$
$$= 1.5708(R^4 - r^4)$$

radius of gyration,

$$k_{AA} = 0.25(D^2 + d^2)^{1/2}$$
$$= 0.5(R^2 + r^2)^{1/2}$$

radius of gyration, polar,

$$k_P = 0.3536(D^2 + d^2)^{1/2}$$
$$= 0.7071(R^2 + r^2)^{1/2}$$

section modulus,

$$Z_{AA} = 0.0982(D^4 - d^4)/D$$
$$Z_P = 0.196(D^4 - d^4)/D$$

CIRCLE, SECTION FORMULAS — *continued*

circle, hollow (semi):

area,

$$A = 0.3927(D^2 - d^2)$$
$$= 1.5708(R^2 - r^2)$$

center of gravity (cg),

$$y = \frac{0.2122(D^3 - d^3)}{D^2 - d^2}$$

$$= \frac{0.4244(R^3 - r^3)}{R^2 - r^2}$$

$$b = \frac{0.5D(D^2 - d^2) - 0.2122(D^3 - d^3)}{D^2 - d^2}$$

diameter,

$$d = 2r$$
$$D = 2R$$

moment of inertia,

$$I_{AA} = 0.0245(D^4 - d^4) - \frac{0.01768(D^3 - d^3)^2}{D^2 - d^2}$$

$$= 0.1098(R^4 - r^4) - \frac{0.283R^2r^2(R - r)}{R + r}$$

radius,

$$r = 0.5d$$
$$R = 0.5D$$

radius of gyration,

$$k_{AA} = (I_{AA}/A)^{1/2}$$

section modulus,

$$Z = I_{AA}/y, \text{ if } y > b$$
$$Z = I_{AA}/b, \text{ if } y < b$$

CIRCLE, SECTION FORMULAS— *continued*

circle sector:

angle,

$$a = 114.6L/D = 57.296L/R$$

area,

$$A = 0.5RL$$
$$= 0.00873aR^2$$
$$= 0.00218aD^2$$

center of gravity (cg),

$$y = 38.2D(\sin 0.5a)/a$$
$$= 76.4R(\sin 0.5a)/a$$

chord length,

$$c = D \sin 0.5a$$
$$= 2R \sin 0.5a$$

length of arc,

$$L = 0.00873aD$$
$$= 0.01745aR$$

moment of inertia,

$$I_{AA} = (0.002182a - 0.25 \sin 0.5a \cos 0.5a)R^4$$
$$I_{BB} = (0.002182a + 0.25 \sin 0.5a \cos 0.5a)R^4$$

radius,

$$R = 0.5D = 57.296L/a$$

radius of gyration,

$$k_{AA} = (I_{AA}/A)^{1/2}$$
$$k_{BB} = (I_{BB}/A)^{1/2}$$

CIRCLE, SECTION FORMULAS — *continued*

circle segment:

angle,

$$a = 114.6L/D = 57.296L/R$$

area,

$$A = 0.5R^2(0.01745a - \sin a)$$

center of gravity (cg),

$$y = \frac{1.333R \sin^3 0.5a}{0.01745a - \sin a}$$

chord length,

$$c = D \sin 0.5a$$
$$= 2R \sin 0.5a$$

height of segment,

$$h = R(1 - \cos 0.5a)$$

length of arc,

$$L = 0.01745aR$$

moment of inertia,

$$I_{AA} = 0.25R^2A + \frac{0.5R^2A \sin^3 0.5a \cos 0.5a}{0.00873a - \sin 0.5a \cos 0.5a}$$

$$I_{BB} = 0.25R^2A - \frac{0.1667R^2A \sin^3 0.5a \cos 0.5a}{0.00873a - \sin 0.5a \cos 0.5a}$$

radius of gyration,

$$k_{AA} = (I_{AA}/A)^{1/2}$$
$$k_{BB} = (I_{BB}/A)^{1/2}$$

---

circle, square inscribed within: SQUARE
circle, triangle inscribed within: TRIANGLES

circle area equal to area of square: SQUARE
CIRCLE FORMULAS: where

a = included angle, degrees
A = area, in.^{2}
c = chord length, in.
C = circumference, in.
cg = center of gravity
D = diameter, in.
h = height of segment. in.
L = length of arc, in.
R = radius, in.
y = dimension, in.

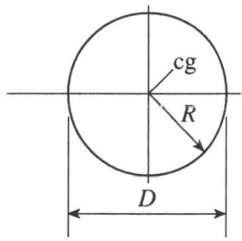

angle a (degrees),

$$a = 114.6L/D = 57.296L/R$$

arc length, sector or segment,

$$L = 0.00873aD = 0.01745aR$$

area, circle,

$$A = 0.7854D^2 = 3.1416R^2$$

area, sector,

$$A = 0.5RL = 0.00218D^2a = 0.00873R^2a$$

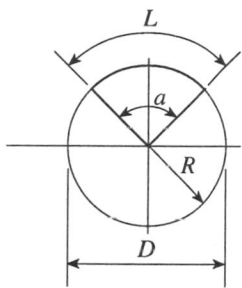

area, segment,

$$A = 0.5R^2(0.01745a - \sin a)$$
$$= 0.5(LR - cy)$$

chord length, segment,

$$c = D \sin 0.5a = 2R \sin 0.5a$$

circumference, circle,

$$C = 3.1416D = 6.2832R$$

diameter,

$$D = 2R = 0.31831C$$

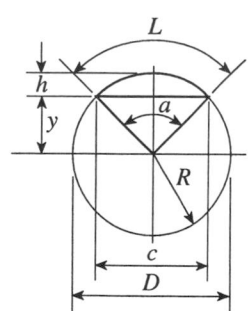

CIRCLE FORMULAS — *continued*

distance (y), segment,

$$y = 0.5(4R^2 - C^2)^{1/2}$$
$$= 0.5(D^2 - C^2)^{1/2}$$
$$= R \cos 0.5a$$

radius,

$$R = 0.5D = 0.1592C$$

segment height,

$$h = R - y = 0.5D(1 - \cos 0.5a)$$
$$= R(1 - \cos 0.5a)$$

---

circle inscribed within triangle: TRIANGLES
circular inch: unit of area equal to the area of a circle 1 in. in diameter.
circular inch =

645.16 circular mm   1 E+06 cmil
5.0671 cm^2 0.7854 in.^{2}

circular mil: unit of area for the cross-section of wires and equal to the area of a circle of 1 mil (0.001 in.) in diameter.
circular mil =

1 E−06 circular in.   0.7854 mil^2
5.0671 E−06 cm^2 5.0671 E−04 mm^2
7.854 E−07 in.^{2}

circular mil-foot: mil-ft, unit of measure per foot of wire of uniform cross-section and area in circular mils.
circular mil-foot =

1.5444 E−04 cm^3
9.425 E−06 in.^{3}

circular millimeter: area of a circle of 1 mm in diameter.

circular millimeter =

0.00155 circular in.	0.001217 in.^{2}
0.007854 cm^2	0.7854 mm^2

circular mil-ohm/foot: OHM-MIL/FOOT
circular motion: ROTATION
Cl: chlorine (element)
Cm: curium (element)
Co: cobalt (element)
coefficient of compressibility of liquids: COMPRESSIBILITY OF VARIOUS LIQUIDS
coefficient of cubical expansion, thermal: HEAT TRANSFER
coefficient of friction: FRICTION
coefficient of heat transfer, conduction: HEAT TRANSFER
coefficient of heat transfer, convection: HEAT TRANSFER
coefficient of heat transfer, overall: HEAT TRANSFER
coefficient of hysteresis: ELECTROMAGNETISM
coefficient of linear expansion, thermal: HEAT TRANSFER
coefficient of thermal conductivity: HEAT TRANSFER
coefficient of viscoscity: VISCOSITY, ABSOLUTE
COLUMN LOADING FORMULAS: where

A = cross-sectional area of column, in.^{2}
F = load, lb
M_{max} = maximum moment, in.-lb
s = compressive stress, lb/in.^{2}
s_W = compressive stress, working, lb/in.^{2}
sf = safety factor, code or design
Z = section modulus, in.^{3}

compressive stress (concentric),

$$s = F/A$$

compressive stress (eccentric),

$$s = F/A + M_{max}/Z$$

COLUMN LOADING FORMULAS — *continued*

compressive stress, working,

$$s_W = s/(\text{sf})$$

COLUMN LOAD LIMITS, based on Euler's formula: where the slenderness ratio is the ratio of the unbraced height (L) of column to its minimum radius of gyration (k), when the slenderness ratio (L/k) is

cast iron with flat ends: >100
cast iron with round ends: >75
oak wood with flat ends: >130
steel with flat ends: >195
steel with hinged ends: >155
steel with round ends: >120
where

c = condition factor
E = modulus of elasticity, lb/in.^{2}
F_A = allowable load, lb
F_M = maximum load, lb
I = moment of inertia, in.^{4}
k = radius of gyration, in.
L = unbraced height of column, in.
r = slenderness ratio
s = stress, maximum, lb/in.^{2}
s_W = stress, working (allowable), lb/in.^{2}
sf = safety factor, code or design

compressive stress, maximum, not to exceed elastic limit of material:

$$s = 9.87cE/r^2$$

compressive stress, working:

$$s_W = s/(\text{sf})$$

compressive stress, working, pipe columns:

$$S_W = 15,200 - 58r$$

COLUMN LOAD LIMITS — *continued*

condition factor (*c*):

both ends fixed: 4
both ends rounded (hinged): 1
one end fixed, one end free: 0.25
one end fixed, one end rounded (hinged): 2
one end fixed, one end guided: 4

load, allowable: $F_A = F_M/(\text{sf})$
load, maximum: $F_M = 9.87cEI/L^2$
slenderness ratio: unbraced height/minimum radius of gyration,

$$r = L/k$$

---

combustion heat value of various materials: HEAT VALUE OF VARIOUS MATERIALS

combustion of fuels, air required: AIR/FUEL RATIO FOR COMBUSTION OF VARIOUS FUELS

common logarithm: LOGARITHMS

compressibility: reciprocal of bulk modulus; the change in volume per unit volume with increase in pressure, where

B = bulk modulus, lb/in.^{2}
k = compressibility coefficient, volume change per pressure change
P_1 = initial pressure, lb/in.^{2}
P_2 = final pressure, lb/in.^{2}
V_1 = initial volume, in.^{3}
V_2 = final volume, in.^{3}

compressibility,

$$k = \frac{1}{B} = \frac{V_2 - V_1}{(P_2 - P_1)V_1}$$

volume change,

$$V_2 - V_1 = kV_1(P_2 - P_1)$$

compressibility coefficient (*k*) of various liquids (1 E−06 per lb/in.^{2}):

carbon disulfide: 4.5   mercury: 0.26

ethyl alcohol: 7.6   oil: 1.36

glycerine: 1.5   water: 3.4

compression/expansion of gases: THERMODYNAMICS

compression strength (ultimate) of various materials: STRENGTH OF MATERIALS

compressors, air/gas: AIR/GAS COMPRESSION FORMULAS

condenser, electrical: ELECTRICAL

conductance, electrical: ELECTRICAL

conductance, heat: HEAT TRANSFER

conductivity, electrical: ELECTRICAL

conductivity, electrical of various materials: ELECTRICAL

CONE FORMULAS: where

$$A = \text{area, ft}^2$$
$$cg = \text{center of gravity}$$
$$d = \text{small diameter, ft}$$
$$D = \text{large diameter, ft}$$
$$h = \text{height, ft}$$
$$I_{AA}, I_{BB}, I_{CC}, I_{DD} = \text{moment of inertia, slug-ft}^2$$
$$k_{AA}, k_{BB}, k_{CC}, k_{DD} = \text{radius of gyration, ft}$$
$$s = \text{slant height, ft}$$
$$V = \text{volume, ft}^3$$
$$W = \text{weight of cone, lb}$$
$$y = \text{dimension, ft}$$

cone, solid:

area of conical surface,

$$A = 1.5708Ds$$

area of total surface,

$$A = 0.7854D^2 + 1.5708Ds$$

center of gravity (cg) through center of cone at

$$y = 0.25h$$

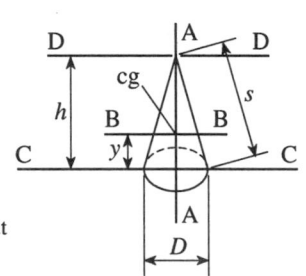

CONE FORMULAS — *continued*

moment of inertia,

$$I_{AA} = 0.0023D^2W$$
$$I_{BB} = 0.00117W(D^2 + h^2)$$
$$I_{CC} = 3.885 \ E-04(3D^2 + 8h^2)W$$
$$I_{DD} = 0.001166(D^2 + 16h^2)W$$

radius of gyration,

$$k_{AA} = 0.27386D$$
$$k_{BB} = 0.19365(D^2 + h^2)^{1/2}$$
$$k_{CC} = 0.1118(3D^2 + 8h^2)^{1/2}$$
$$k_{DD} = 0.19365(D^2 + 16h^2)^{1/2}$$

slant height,

$$s = (0.25D^2 + h^2)^{1/2}$$

volume,

$$V = 0.2618D^2h$$

cone frustrum, solid:

area of conical surface,

$$A = 1.5708(D + d)s$$

center of gravity (cg) through center of cone at

$$y = \frac{0.25h(D^2 + 2Dd + 3d^2)}{D^2 + Dd + d^2}$$

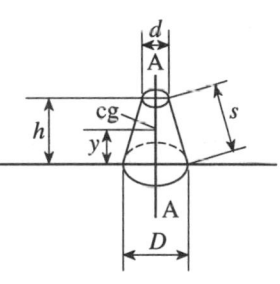

moment of inertia,

$$I_{AA} = \frac{0.0023W(D^5 - d^5)}{D^3 - d^3}$$

radius of gyration,

$$k_{AA} = \frac{0.27386(D^5 - d^5)^{1/2}}{(D^3 - d^3)^{1/2}}$$

CONE FORMULAS — *continued*

slant height,

$$s = [h^2 + 0.25(D - d)^2]^{1/2}$$

volume,

$$V = 0.2618h(D^2 + Dd + d^2)$$

convection, heat: HEAT TRANSFER

conversion of units: replace unit to be converted with the equivalent of the desired unit, as

mi/h to ft/s
replace mi with (=) 5280 ft
h with (=) 3600 s
mi/h = 5280 ft/3600 s
= 1.4667 ft/s

Copper Wire Gauges (American Wire Gauge)

Size		Dia. (in.)	Electrical resistance (ohms/1000 ft)	Weight (lb/1000 ft)
Solid	Stranded			
50		0.0010	10800	0.003
48		0.0012	7200	0.004
46		0.0016	4051	0.008
44		0.0020	2590	0.012
42		0.0025	1660	0.019
40		0.0031	1079	0.029
39		0.0035	847	0.037
38		0.0040	648	0.048
37		0.0045	512	0.061
36		0.0050	415	0.076
35		0.0056	331	0.095
34		0.0063	261	0.120
33		0.0071	206	0.153

Copper Wire Gauges — *continued*

Size		Dia. (in.)	Electrical resistance (ohms/1000 ft)	Weight (lb/1000 ft)
Solid	Stranded			
32		0.0080	162	0.194
31		0.0089	131	0.240
30		0.0100	104	0.303
29		0.0113	81.2	0.387
28		0.0126	65.3	0.481
27		0.0142	51.4	0.610
26		0.0159	41.0	0.765
25		0.0179	32.4	0.970
24		0.0201	25.7	1.22
23		0.0226	20.3	1.55
22		0.0254	16.2	1.95
21		0.0285	12.8	2.46
20		0.0320	10.1	3.10
	20	0.0360	10.1	3.16
19		0.0359	8.05	3.90
18		0.0403	6.39	4.92
	18	0.0460	6.41	5.01
17		0.0453	5.05	6.21
16		0.0508	4.02	7.81
	16	0.0580	4.02	7.98
15		0.0571	3.18	9.87
14		0.0641	2.52	12.40
	14	0.0730	2.53	12.70
13		0.0720	2.00	15.70
12		0.0808	1.59	19.80
	12	0.0920	1.59	20.20
11		0.0907	1.26	24.90
10		0.1019	0.999	31.43
	10	0.1160	1.00	32.00
9		0.1144	0.7925	39.61
	9	0.1300	0.7939	40.40

Copper Wire Gauges — *continued*

Size		Dia. (in.)	Electrical resistance (ohms/1000 ft)	Weight (lb/1000 ft)
Solid	Stranded			
8		0.1285	0.6281	49.98
	8	0.1460	0.6273	51.00
7		0.144	0.5002	63
	7	0.164	0.4988	64
6		0.162	0.3952	79
	6	0.184	0.3956	81
5		0.182	0.3135	101
	5	0.206	0.3130	102
4		0.204	0.2485	126
	4	0.232	0.2486	129
3		0.229	0.1976	159
	3	0.260	0.1971	163
2		0.258	0.1539	201
	2	0.292	0.1562	205
1		0.289	0.1239	253
	1	0.332	0.1238	258
0		0.325	0.0982	320
	0	0.373	0.0984	326
00		0.365	0.0793	403
	00	0.419	0.0779	411
000		0.410	0.0168	508
	000	0.470	0.0618	518
0000		0.460	0.0491	640
	0000	0.528	0.0490	653
	250 mcm	0.575	0.0415	772
	300 mcm	0.630	0.0346	926
	350 mcm	0.681	0.0296	1081
	400 mcm	0.728	0.0259	1235
	450 mcm	0.772	0.0231	1389
	500 mcm	0.813	0.0208	1540
	550 mcm	0.853	0.0187	1698

Copper Wire Gauges — *continued*

Size		Dia. (in.)	Electrical resistance (ohms/1000 ft)	Weight (lb/1000 ft)
Solid	Stranded			
	600 mcm	0.891	0.0173	1853
	650 mcm	0.929	0.0160	2007
	700 mcm	0.964	0.0148	2161
	750 mcm	0.998	0.0138	2316
	800 mcm	1.031	0.0130	2470
	850 mcm	1.062	0.0122	2625
	900 mcm	1.094	0.0115	2779
	950 mcm	1.122	0.0109	2935
	1000 mcm	1.152	0.0106	3088
	1100 mcm	1.209	0.0093	3400
	1200 mcm	1.263	0.0087	3710
	1250 mcm	1.289	0.0083	3859
	1300 mcm	1.315	0.0080	4010
	1400 mcm	1.364	0.0074	4320
	1500 mcm	1.412	0.0069	4631
	1600 mcm	1.459	0.0065	4940
	1700 mcm	1.504	0.0061	5250
	1750 mcm	1.526	0.0059	5403
	1800 mcm	1.548	0.0058	5560
	1900 mcm	1.590	0.0055	5870
	2000 mcm	1.632	0.0053	6175
	2500 mcm	1.824	0.0042	7794
	3000 mcm	1.998	0.0035	9353
	3500 mcm	2.159	0.0031	11018
	4000 mcm	2.309	0.0026	12592
	4500 mcm	2.448	0.0024	14303
	5000 mcm	2.581	0.0021	15892

cord: unit of measure for a pile of wood.
cord =

8 cord-ft	3.625 m^3
8 ft × 4 ft × 4 ft	3.625 steres
128 ft^3	

cord-foot =

0.125 cord	16 ft^3
4 ft × 4 ft × 1 ft	

cosecant; csc: TRIGONOMETRY
cosine; cos: TRIGONOMETRY
cosine law: TRIGONOMETRY
cotangent; cot: TRIGONOMETRY
coulomb: ELECTROSTATICS
coulomb =

0.1 abcoulomb	1.036 E−05 faraday
2.778 E−04 A-h	1 E+06 μC
1 A-s	3 E+09 statcoulombs
6.2832 E+18 electronic charges	

coulomb/second: AMPERE
coulomb/square centimeter =

0.1 abcoulomb/cm^2	
6.4516 C/in.^{2}	

coulomb/square inch =

0.0155 abcoulomb/cm^2	1550 C/m^2
0.1550 C/cm^2	4.65 E+08 statcoulombs/cm^2

coulomb/square meter =

1 E−04 C/cm^2	
6.452 E−04 C/in.^{2}	

Coulomb's law: ELECTROSTATICS, MAGNETISM
couple: STATICS
Cr: chromium (element)
Cs: cesium (element)
Cu: copper (element)
CUBE, FORMULAS: prism with 6 equal sides, where

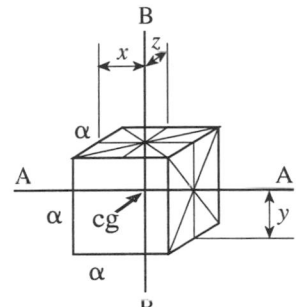

a, x, y, z = dimensions, ft
A = area, ft^2
I_{AA}, I_{BB} = moment of inertia, slug-ft^2
k_{AA}, k_{BB} = radius of gyration, ft
V = volume, ft^3
W = weight, lb

cube, solid:

area of surface,

$$A = 6a^2$$

center of gravity, through center of cube,

$$x = y = z = 0.5a$$

moment of inertia,

$$I_{AA} = I_{BB} = 0.0052a^2 W$$

radius of gyration,

$$k_{AA} = k_{BB} = 0.408a$$

volume,

$$V = a^3$$

---

cubical expansion, thermal: HEAT TRANSFER
cubic centimeter =

4.238 E−04 board-ft	0.004227 cup
2.838 E−05 bushel (level)	1 E−04 dal
0.1 cl	1 E−09 dam^3

cubic centimeter — *continued*

0.001 dm³	6475 mil-ft
0.2705 dr (liquid)	16.23 minims
3.53 E−05 ft³	1 ml
2.270 E−04 gal (dry)	1000 mm³
2.2 E−04 gal (Imperial)	0.033814 oz (liquid)
2.642 E−04 gal (Liquid)	1.135 E−04 peck
0.008454 gill	0.001816 pt (dry)
1 E−05 hl	0.002113 pt (liquid)
0.06102 in.³	9.081 E−04 qt (dry)
1 E−06 kl	0.001057 qt (liquid)
0.001 l	0.0676 tbsp
0.0022 lb of water	0.203 tsp
1 E−06 m³	1.308 E−06 yd³

cubic centimeter/gram = 0.01602 ft³/lb
cubic centimeter/second =

60 cm³/min	0.01585 gal/min
0.127 ft³/h	2.642 E−04 gal/s
0.002119 ft³/min	3.66 in.³/min
3.5315 E−05 ft³/s	1 ml/s

cubic centimeter-atmosphere =

9.6102 E−05 Btu	0.024222 g-cal
1.0133 E+06 erg	0.101325 J
3.5315 E−05 ft³-atm	2.422 E−05 kg-cal
2.4045 ft-pdl	2.815 E−05 Wh

cubic decameter: CUBIC DEKAMETER
cubic decimeter: LITER
cubic dekameter =

1 E+09 cm³	6.1024 E+07 in.³
1 dam³	1 E+06 l
1 E+06 dm³	1000 m³
3.5315 E+04 ft³	

cubic foot =

2.296 E − 05 acre-ft	0.11874 hogshead
2.755 E − 04 acre-in.	1728 in.^{3}
0.2449 bbl (dry)	0.028317 kl
0.2375 bbl (liquid)	28.317 l
0.1781 bbl (petroleum)	0.028317 m^3
12 board-ft	2.8317 E + 04 ml
0.8036 bushel (level)	957.5 oz (liquid)
2.8317 E + 04 cm^3	3.214 pecks
0.007813 cord (wood)	0.0404 perch (masonry)
0.0625 cord-ft (wood)	51.428 pt (dry)
2.8317 E − 05 dam^3	59.84 pt (liquid)
28.317 dm^3	25.7136 qt (dry)
6.4285 gal (dry)	29.922 qt (liquid)
6.229 gal (Imperial)	0.028317 stere
7.4805 gal (liquid)	0.03704 yd^3
0.28316 hl	

cubic foot/hour =

2.296 E − 05 acre-ft/h	7.4805 gal/h
471.947 cm^3/min	0.1247 gal/min
7.866 cm^3/s	28.8 in.^{3}/min
0.01667 ft^3/min	28.317 l/h
2.7778 E − 04 ft^3/s	471.95 ml/min
	7.866 ml/s

cubic foot/minute =

0.001377 acre-ft/h	0.1247 gal/s
2.296 E − 05 acre-ft/min	28.317 l/min
471.95 cm^3/s	0.4719 l/s
60 ft^3/h	1.699 m^3/h
0.0167 ft^3/s	0.028317 m^3/min
448.8 gal/h	4.719 E − 04 m^3/s
7.4805 gal/min	471.96 ml/s

cubic foot/pound =

  62.438 cm^3/g
  0.06244 m^3/kg

cubic foot/second =

0.9917 acre-in./h	7.4805 gal/s
2.8317 E + 04 cm^3/s	1699 l/min
3600 ft^3/h	28.317 l/s
60 ft^3/min	1.699 m^3/min
6.46317 E + 05 gal/d	0.028317 m^3/s
2.693 E + 04 gal/h	2.8317 E + 04 ml/s
448.831 gal/min	2.222 yd^3/min

cubic foot-atmosphere =

2.7213 Btu	0.0010689 hp-h
2.8317 E + 04 cm^3-atm	2869.2 J
2.869 E + 10 erg	292.578 kg-m
2116.224 ft-lb	7.9705 E − 04 kWh
6.8087 E + 04 ft-pdl	28.317 l-atm
685.76 g-cal	

cubic foot-atmosphere/hour = 7.9705 E − 04 kW
cubic foot-atmosphere/pound =

  2.7213 Btu/lb
  1.757 E − 06 kWh/g

cubic inch =

1.417 E − 04 bbl (dry)	0.016387 dm^3
1.374 E − 04 bbl (liquid)	4.333 dr (liquid)
0.006944 board-ft	5.787 E − 04 ft^3
4.651 E − 04 bushel (level)	0.00372 gal (dry)
1.6387 cl	0.003606 gal (Imperial)
16.387 cm^3	0.004329 gal (liquid)
0.0693 cup	0.13853 gill
1.64 E − 08 dam^3	6.87 E − 05 hogshead

cubic inch — *continued*

1.639 E−05 kl	0.00186 peck
0.01639 l	0.02976 pt (dry)
1.6387 E−05 m³	0.03463 pt (liquid)
1.061 E+05 mil-ft	0.01488 qt (dry)
266 minims	0.01732 qt (liquid)
16.387 ml	1.108 tbsp
1.6387 E+04 mm³	3.333 tsp
0.5541 oz (liquid)	2.143 E−05 yd³

cubic inch/minute =

16.387 cm³/min	0.0347 ft³/h
0.273 cm³/s	2.7312 E−07 m³/s

cubic meter =

8.107 E−04 acre-ft	220.1 gal (Imperial)
0.009728 acre-in.	264.172 gal (liquid)
8.648 bbl (dry)	10 hl
8.386 bbl (liquid)	6.1024 E+04 in.³
6.2898 bbl (petroleum)	1 kl
28.38 bushels (level)	1000 l
1 E+06 cm³	1 E+09 mm³
0.001 dam³	3.3814 E+04 oz (liquid)
10 decisteres	2113 pt (liquid)
0.1 dekastere	1056.7 qt (liquid)
1000 dm³	1 stere
35.3145 ft³	1.308 yd³

cubic meter/hour =

0.5886 ft³/min	0.2778 l/s
4.403 gal/min	0.01667 m³/min
16.667 l/min	2.778 E−04 m³/s

cubic meter/kilogram = 16 ft³/lb

cubic meter/minute =

35.315 ft³/min	1000 l/min
0.5886 ft³/s	60 m³/h
1.585 E+04 gal/h	0.01667 m³/s
264.172 gal/min	

cubic meter/second =

2118.9 ft³/min	3600 m³/h
1.585 E+04 gal/min	60 m³/min
6 E+04 l/min	

cubic millimeter =

0.001 cm³	1 E−09 m³
6.1024 E−05 in.³	0.016321 minim

cubic yard =

6.1983 E−04 acre-ft	202 gal (liquid)
21.696 bushels (level)	46,656 in.³
7.646 E+05 cm³	0.7646 kl
7.646 E−04 dam³	764.6 l
764.56 dm³	0.7646 m³
27 ft³	1388.56 pt (dry)
173.569 gal (dry)	1615.79 pt (liquid)
168.18 gal (Imperial)	807.9 qt (liquid)

cubic yard/minute =

0.45 ft³/s	12.74 l/s
12,118 gal/h	0.01274 m³/s
3.366 gal/s	

cubit =

45.72 cm	18 in.
1.5 ft	0.5 yd

cup =

236.5 cm^3	16 tbsp
14.43 in.^{3}	48 tsp
8 oz (liquid)	

current: ELECTRICAL

cycle = 0.001 kc

cycle/second =

1 Hz	1 E−06 Mc/s
0.001 kc/s	

cycloid: a curve generated by a point on the circumference of a moving circle, where

A = area, in.^{2}
D = diameter of circle, in.
s = length of arc, in.

area of cycloid:

$A = 2.356D^2$

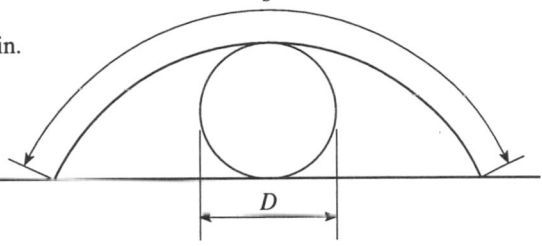

length of arc generated, per revolution:

$$s = 4D$$

CYLINDER FORMULAS: where

A = area, ft^2
cg = center of gravity
d = inside diameter, ft
D = outside diameter, ft
I_{AA}, I_{BB}, I_{CC} = moment of inertia, slug-ft^2
k_{AA}, k_{BB}, k_{CC} = radius of gyration, ft
L = length of cylinder, ft
V = volume, ft^3
W = weight, lb
y = dimension, ft

CYLINDER FORMULAS — *continued*

cylinder, solid:

area of each cylinder end,

$$A = 0.7854D^2$$

area of cylinder surface,

$$A = 3.1416DL$$

area, total surface of cylinder,

$$A = 1.5708D^2 + 3.1416DL$$

center of gravity (cg), through center of cylinder, at

$$y = 0.5L$$

moment of inertia,

$$I_{AA} = 0.00389D^2W$$
$$I_{BB} = 0.00259(0.75D^2 + L^2)W$$
$$I_{CC} = 0.00259(0.75D^2 + 4L^2)W$$

radius of gyration,

$$k_{AA} = 0.3536D$$
$$k_{BB} = 0.2887(0.75D^2 + L^2)^{1/2}$$
$$k_{CC} = 0.2887(0.75D^2 + 4L^2)^{1/2}$$

volume,

$$V = 0.7854D^2L$$

cylinder, thick hollow:

area of each cylinder end,

$$A = 0.7854(D^2 - d^2)$$

area of cylinder inside surface,

$$A = 3.1416dL$$

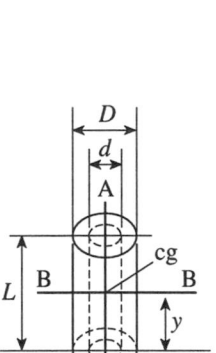

CYLINDER FORMULAS — *continued*

area of cylinder outside surface,

$$A = 3.1416DL$$

area, total surface of cylinder,

$$A = 1.5708(D^2 - d^2) + 3.1416(D + d)L$$

center of gravity (cg), through center of cylinder, at

$$y = 0.5L$$

moment of inertia,

$$I_{AA} = 0.00389(D^2 + d^2)W$$
$$I_{BB} = 6.48 \text{ E}-04 \ (3D^2 + 3d^2 + 4L^2)W$$

radius of gyration,

$$k_{AA} = 0.3536(D^2 + d^2)^{1/2}$$
$$k_{BB} = 0.1443(3D^2 + 3d^2 + 4L^2)^{1/2}$$

volume,

$$V = 0.7854(D^2 - d^2)L$$

cylinder, thin hollow:

area of cylinder surface,

$$A = 3.1416DL$$

center of gravity (cg), through center of cylinder, at

$$y = 0.5L$$

moment of inertia,

$$I_{AA} = 0.0078D^2W$$
$$I_{BB} = 0.0013(3D^2 + 2L^2)W$$

radius of gryration,

$$k_{AA} = 0.5D$$
$$k_{BB} = 0.2887(1.5D^2 + L^2)^{1/2}$$

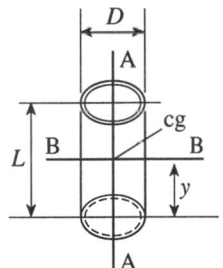

D

Dalton's gas law: GAS LAWS
daraf: ELECTROSTATICS
datum: level of reference for height measurement
day =

24 h	0.032877 month
1440 min	86,400 s

deca: DEKA
decagram: DEKAGRAM
decameter: DEKAMETER
deci (d): prefix, equals 0.1
decibel (dB): SOUND
decibel =

0.1 B
0.115128 Np

decigram =

10 cg	1 E−04 kg
0.1 g	

deciliter =

10 cl
0.1 l

decimeter =

10 cm	0.1 m
0.3281 ft	100 mm
0.001 hm	0.1094 yd
3.937 in.	

decistere =

0.1 m³
0.1 stere

degC: degrees Celsius, temperature (formerly degrees Centigrade).
degC = 1.8°F
         = K − 273.16
degC absolute: K
degC convert from °F:

$$°C = 0.5556(°F − 32)$$

degF: degrees Fahrenheit, temperature

degF = 0.5556°C
         − °R − 459.69

degF absolute: °R
degF convert from °C:

$$°F = (1.8 × °C) + 32$$

degK: Kelvin, °C (absolute)

degK = °C + 273.16
         = 0.5556 × °R

degK convert from °F:

$$K = 0.5556(°F + 459.6)$$

degR: degrees Rankine, °F (absolute)

⑧ degR = °F + 459.69
          = 1.8 K

degree (angular): measure of a plane angle, in degrees, as part of a 360° circle.

degree (angular) =

0.00278 of circle	0.017453 rad
60 min	0.00278 rev
0.01111 quadrant	3600 s

degree (angular)/inch = 0.006871 rad/cm

degree (angular)/minute =

0.01667°/s	4.63 E−05 rev/s
2.909 E−04 rad/s	

degree (angular)/second =

60°/min	0.1667 rev/min
0.017453 rad/s	0.002778 rev/s

degree day: term used to evaluate space load requirements, as measured by the difference between the average temperature for a day (24-h period) and 65°F. Degree days are totaled for season heating and cooling equipment requirements.

degrees API: API gravity; petroleum industry system for relating specific gravity of lubricating oils, measured in degrees, where sp gr is that of the oil at 60°F.

$$degrees\ API = 141.5/(sp\ gr) - 131.5$$

degrees API of various liquids:

fuel oil no. 1: 38	fuel oil no. 5: 20
no. 2: 34	no. 6: 14
no. 3: 30	kerosene: 42
no. 4: 26	

degrees Baumé: scale system for relating specific gravities of liquids measured in degrees, per National Institute of Standards and Technology (formerly U.S. Bureau of Standards), based on water as standard at 10° Baumé, where R = specific gravity of the oil at 60°F.

$$\text{degrees Baumé} = 140/R - 130$$

deka (da): prefix, equals 10

dekagram =

1 dag	0.32151 oz (apoth/troy)
10 g	0.35274 oz (avdp)
0.1 hg	

dekaliter =

10 l	18.16217 pt (dry)
1.135 pecks	

dekameter =

1000 cm	393.701 in.
1 dam	0.01 km
100 dm	10 m
32.8084 ft	1 E + 04 mm
2.6417 gal	1.9884 rods
0.1 hm	10.93613 yd

dekastere =

1 E + 04 l	10 steres
10 m³	

density, absolute: mass per unit volume

English system of units, specific weight divided by gravitational acceleration, where

$$d = \text{density, slugs/ft}^3$$
$$g = \text{gravitational acceleration, 32.174 ft/s}^2$$
$$w = \text{specific weight, lb/ft}^3$$

density,

$$d = w/g = 0.0311w$$

Metric system of units, density in gram/cm³

dewpoint temperature of air: temperature at which saturation of the air with moisture (water) is complete, whereas at a temperature above this point, water will vaporize and below this point water will condense.

dielectric: ELECTROSTATICS

dielectric constant: ELECTROSTATICS

dielectric heating: HEAT TRANSFER

dielectric strength: ELECTROSTATICS

diethyl ether: ether

DISK FORMULAS: where

$$A = \text{area, ft}^2$$
$$D, t = \text{dimensions, ft}$$
$$I_{AA}, I_{BB} = \text{moment of inertia, slug-ft}^2$$
$$k_{AA}, k_{BB} = \text{radius of gyration, ft}$$
$$V = \text{volume, ft}^3$$
$$W = \text{weight, lb}$$

disk, solid:

area of total surface,

$$A = 1.571D^2 + 3.142Dt$$

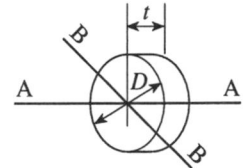

DISK, FORMULAS — *continued*

moment of inertia,

$$I_{AA} = 0.00389D^2W$$
$$I_{BB} = 0.00194D^2W$$

radius of gyration,

$$k_{AA} = 0.3536D$$
$$k_{BB} = 0.25D$$

volume,

$$V = 0.7854D^2t$$

displacement: distance, position change of an object from initial to final position, as in angular or linear displacement.

dozen: 12 items

drachm: dram

dram =

0.4557 dr (apoth/troy)	0.004747 lb (apoth/troy)
1 dr (avdp)	1772 mg
1.772 g	0.0625 oz
27.344 gr	0.05697 oz (apoth/troy)
0.001772 kg	1.139 pwt
0.003906 lb	1.367 scruples (apoth/troy)

dram (apoth/troy) =

2.1943 dr (avdp)	3888 mg
3.888 g	0.125 oz
60 gr	0.13714 oz (avdp)
0.00389 kg	2.5 pwt
0.010417 lb	3 scruples
0.00857 lb (avdp)	

dram (avdp): DRAM

dram (fluid): DRAM (LIQUID)

dram (liquid) =

3.697 cm³	3.697 ml
0.03125 gill	0.125 oz (liquid)
0.2256 in.³	0.007813 pt (liquid)
3.6967 E−06 kl	0.003906 qt (liquid)
0.0037 l	0.25 tbsp
60 minims	0.75 tsp

dry bulb temperature: actual temperature of air as read with a standard thermometer.

Dy: dysprosium (element)

dynamics: study of motion and forces causing the motion, as in LINEAR MOTION and ROTATION

dynamic viscosity: VISCOSITY, ABSOLUTE

dyne, dyn: metric unit of force that accelerates 1 gram of mass, 1 cm/s².

dyne =

0.0010197 g	2.248 E−06 lb (force)
0.015737 gr	1.01972 mg
1 E−07 J/cm	1 E−05 N
1 E−05 J/m	7.233 E−05 pdl
1.02 E−06 kg	1.124 E−09 ton

dyne/centimeter =

1 erg/cm²	1.0197 mg/cm
0.01 erg/mm²	2.59 mg/in.
0.0010197 g/cm	0.10197 mg/mm
5.71 E−06 lb (force)/in.	1.837 E−04 pdl/in.

dyne/cubic centimeter =

0.0010197 g/cm³
0.001185 pdl/in.³

dyne/inch = 1.01972 mg/in.

dyne/square centimeter =

9.869 E − 07 atm	0.002088 lb/ft^2
1 E − 06 bar	1.450 E − 05 lb/in.^{2}
1 barye	1 μbar
7.5 E − 05 cm (mercury, 0°C, 32°F)	0.750 μm (mercury, 0°C, 32°F)
0.0010197 cm (water, 4°C, 39.2°F)	0.001 mbar
3.3456 E − 05 ft (water, 4°C, 39.2°F)	7.5 E − 04 mm (mercury, 0°C, 32°F)
0.0010197 g/cm^2	0.10 N/m^2
2.953 E − 05 in. (mercury, 0°C, 32°F)	0.10 Pa
4.0147 E − 04 in. (water, 4°C, 39.2°F)	4.666 E − 04 pdl/in.^{2}
1.0197 E − 06 kg/cm^2	1.044 E − 06 ton/ft^2
0.010197 kg/m^2	7.252 E − 09 ton/in.^{2}
1 E − 04 kPa	7.5006 E − 04 torr

dyne-centimeter: ERG
dyne-centimeter/second: ERG/SECOND
dyne-second/square centimeter: POISE

E

e: logarithmic constant equals 2.7182818285
Earth:

  density =

  5.522 g/cm^3 345 lb/ft^3
  5522 kg/m^3

  diameter =

  4.186 E+07 ft   7928.7 m
  12,760 km

  mass =

  6.019 E+24 kg
  1.327 E+25 lb

  volume =

  3.85 E+22 ft^3
  1.09 E+21 m^3

eddy current: ELECTROMAGNETISM
efficiency (eff) = actual/theoretical = output/input
efficiency, percent = 100 (efficiency in decimals)
                       = 100 (actual/theoretical)
                       = 100 (output/input)

elastic limit: STRENGTH OF MATERIALS
ELECTRICAL: where

a = electrical resistivity coefficient for
   temperature correction for a material, per/°F
A = cross-sectional area, in.^{2}
AC = alternating current
B = angle, degree
C = capacitance, farads
C_1, C_2, C_3 = individual capacitance, farads
C_F = capacitance of cable, microfarads/ft
C_M = capacitance of cable, microfarads/mi
C_T = total capacitance, farads
cm = area of conductor, cmil
d = length of conductor, ft
d_1 = length of circuit, one way, ft
D = diameter of conductor, in.
D_C = outside diameter of cable insulation, in.
D_W = diameter of wire, in.
DC = direct current
E = potential, volts
E_{AO}, E_{BO}, E_{CO} = phase to common potential, volts
E_P = transformer primary winding potential, volts
E_S = transformer secondary winding potential, volts
eff = efficiency, decimal
emf = electromotive force, volts
f = frequency, Hz (cycle/sec)
g = conductivity of a material, mho-ft/mil
G = conductance, mhos
G_1, G_2, G_3 = individual conductances, mhos
G_T = total conductance, mhos
H = energy, joules
hp = horsepower
Hz = hertz, frequency (c/sec)
i = current change, ampere/sec
I = current, amperes
I_A, I_B, I_C = phase current, amperes
I_P = transformer primary winding current, amperes
I_S = transformer secondary winding current, amperes
k = dielectric constant of insulation

ELECTRICAL — *continued*

kW = power, kilowatts
kWh = energy, kilowatt-hour
kVA = power, kilovolt-ampere
$kVAR$ = reactive power, kilovars
l_0 = neutral line
l_1, l_2, l_3 = power lines
L = inductance, henrys
LRA = locked rotor amperes
m = magnetic flux change, maxwell/sec
N = speed, rev/min
N_S = synchronous speed, rev/min
p = number of poles of a motor
pf = power factor
Q = capacitor, charge, Coulombs
r = resistivity of material, ohm-mil/ft
r_1 = resistivity of a material at known temperature, ohm-mil/ft
r_2 = resistivity of a material at another temperature, ohm-mil/ft
R = resistance, ohms
R_1, R_2, R_3 = individual resistances, ohms
R_T = total resistance, ohms
s = space between centers of cables, in.
S = slip, percent
t_1 = temperature at known resistivity, °F
t_2 = temperature at another resistivity, °F
t_m = time, seconds
T = torque, ft-lb
T_C = number of turns in a coil
T_P = number of turns in transformer primary winding
T_S = number of turns in transformer secondary winding
VA = power, volt-amperes
VD = voltage drop, volts
W = power, watts
WH = energy, watt-hours
X = total reactance, ohms
X_C = capacitive reactance, ohms
X_L = inductive reactance, ohms
Z = total impedance, ohms

ELECTRICAL — *continued*

absolute system of units: International electrical units   *

aluminum wire gauges, resistances and weights: ALUMINUM WIRE GAUGES

ampere: unit for quantity of electrical current that flows through a resistance of 1 ohm when the electromotive force is 1 volt. One ampere of electrical current passing through a solution of silver nitrate deposits 0.001118 gram of silver per second.

apparent power: actual power flowing in an AC electrical system, in kVA or VA.

autotransformer: type of electric transformer utilizing a single coil with two primary (input voltage) leads and two secondary (output voltage) leads. Secondary leads are tapped at points of voltage desired.

capacitance (*C*): term for the charge that is stored in an electrical circuit, measured in farads. The charge increases as the voltage rises, holds when voltage is constant, and decreases as the voltage falls. Capacitance opposes voltage changes in an electrical circuit.

capacitance charge in a circuit,

$$Q = CE$$

capacitance total, of capacitors connected in parallel,

$$C_T = C_1 + C_2 + C_3 + \cdots + C_n$$

capacitance total, of capacitors connected in series,

$$1/C_T = 1/C_1 + 1/C_2 + 1/C_3 + \cdots + 1/C_n$$

capacitance of electrical cables:

insulated cable, capacitance,

$$C_F = 7.35 \, \text{E} - 06 \, k/\log (D_C/D_W)$$

parallel cables,

$$C_M = 0.01941 k/\log (2s/d)$$

three-phase balanced AC power line, in a triangular configuration (phase to neutral),

$$C_M = 0.03882 k/\log (2s/d)$$

ELECTRICAL — *continued*

capacitive reactance: reactance, capacitive *

capacitor, condenser: an electrical device consisting of two plates (electrodes) separated by an insulating material (dielectric) and used for storing electrical charges in an electrical circuit when a voltage is applied to the plates.

condenser: capacitor *

conductance: ability to transmit electrical current, measured in mhos and is the reciprocal of resistance.

conductance,

$$G = 1/R$$

conductance of a conductor,

$$G = g(\text{cm})/d$$
$$= 1\ \text{E}+06\ gD^2/d$$
$$= 1.27\ \text{E}+06\ gA/d$$

conductance total, of conductors connected in parallel,

$$G_T = G_1 + G_2 + G_3 + \cdots + G_n$$

conductivity: specific conductance of a material measured as conductance per cubic centimeter (mho/cm^3), or conductance per circular mil per foot (mho-ft/mil). Also may be measured as percent of copper, where copper is 580,000 mho/cm, or 0.09642 mho-ft/mil.

conductivity (*g*) of various materials (as percent copper) (copper = 0.09642 mho-ft/mil):

aluminum, annealed: 62	bronze (0.10 zinc): 45
hard drawn: 61	bronze/phosphor: 38
pure: 64	cadmium: 25
antimony: 4.1	calcium: 37
arsenic: 4.8	cerium: 2.2
barium: 18	cesium: 9.1
beryllium: 9.3	chromium: 62
bismuth: 1.4	cobalt: 18
brass, annealed: 25	columbium: 14.4
yellow: 28	constantan: 3.5

ELECTRICAL — *continued*

COPPER, ANNEALED: 100
    hard drawn: 96
germanium: 3.7
german silver: 4.2
gold: 71
graphite: 0.2
hafnium: 4.5
inconel: 1.7
indium: 21
invar: 2
iridium: 32.5
iron, cast: 2.5
    malleable: 10
    pure: 18
    wrought: 12
lead: 7.8
lithium: 20
magnesium: 39
manganese: 0.9
manganin: 3.9
mercury: 1.8
molybdenum: 31
monel: 3.6
nichrome: 1.7
nickel: 22
niobium: 14

osmium: 18
palladium: 16
platinum: 17.5
potassium: 25
rhodium: 31
ruthenium: 22.7
selenium: 14.4
silver: 105
sodium: 32.8
steel, carbon: 7.5
    silicon: 3.4
    soft: 11
    stainless: 2.4
    structural: 12
strontium: 6.9
tantalum: 13.9
thallium: 9.8
thorium: 9.6
tin: 15
titanium: 3.6
tungsten: 31
uranium: 5.9
vanadium: 6.7
zinc: 30
zirconium: 4.1

copper wire gauges, resistances and weights: COPPER WIRE GAUGES
coulomb: ELECTROSTATICS
Coulomb's law: ELECTROSTATICS
current: rate of flow of electricity in a circuit, measured in amperes.
 **current, AC circuits:**

capacitive reactance only,

$$I = E/X_C$$

capacitive reactance, inductive reactance and resistance (in parallel),

$$I = [(E/R)^2 + (E/X_L - E/X_C)^2]^{1/2}$$
$$= [(E/R)^2 + (E/X)^2]^{1/2}$$

ELECTRICAL — *continued*

capacitive reactance, inductive reactance and resistance (in series),

$$I = \frac{E}{[R^2 + (X_L - X_C)^2]^{1/2}}$$

$$= \frac{E}{(R^2 + X^2)^{1/2}}$$

capacitive reactance and resistance (in parallel),

$$I = [(E/R)^2 + (E/X_C)^2]^{1/2}$$

capacitive reactance and resistance (in series),

$$I = \frac{E}{[R^2 + (X_C)^2]^{1/2}}$$

impedance,

$$I = E/Z$$

inductive reactance only,

$$I = E/X_L$$

inductive reactance and resistance (in parallel),

$$I = [(E/R)^2 + (E/X_L)^2]^{1/2}$$

inductive reactance and resistance (in series),

$$I = \frac{E}{[R^2 + (X_L)^2]^{1/2}}$$

resistance only,

$$I = E/R$$

single-phase circuit, 2-wire,

$$I = \frac{1000 \text{ kW}}{E(\text{pf})}$$

$$= 1000 \text{ kVA}/E$$

ELECTRICAL — *continued*

single-phase circuit, 3-wire,

$$I = \frac{500 \ \text{kW}}{E_{AO}(\text{pf})}$$

single-phase motor,

$$I = \frac{746 \ \text{hp}}{E(\text{eff})(\text{pf})}$$

single-phase transformer,

$$I = 1000 \ \text{kVA}/E$$

three-phase circuit, 3-wire, delta-connected,

$$I = \frac{1000 \ \text{kW}}{1.732E(\text{pf})}$$

$$= \frac{1000 \ \text{kVA}}{1.732E}$$

balanced load,

$$I = 1.732I_A = 1.732I_B = 1.732I_C$$

three-phase circuit, 4-wire, wye-connected,

$$I_A = \frac{1000 \ \text{kW}}{3E_{AO}(\text{pf})}$$

(I_B, E_{BO} and I_C, E_{CO} are similar)
balanced load,

$$I = I_A = I_B = I_C$$

three-phase motor,

$$I = \frac{746 \times \text{hp}}{1.732 \times E \times \text{eff} \times \text{pf}}$$

three-phase transformer,

$$I = \frac{1000 \ \text{kVA}}{1.732 \times E}$$

ELECTRICAL — *continued*

Ⓐ **current, dc circuits:**

$$I = E/R = (1000 \times kW)/E$$
$$= VA/E = (VA/R)^{1/2}$$

motor,

$$I = \frac{746 \times hp}{E \times eff}$$

dielectric: ELECTROSTATICS
dielectric constants: ELECTROSTATICS
efficiency, eff:

$$eff = output/input$$

electromagnetism: ELECTROMAGNETISM
electromotive force, emf: volt   *
farad: unit of measurement for capacitance and is equal to the transfer of 1 C (ampere/sec) of electricity per each volt of difference in potential between two conductors. Also the measure of capacitance of a capacitor (condenser) charged to a potential of 1 V by 1 ampere/sec.
frequency: number of cycles per second in an AC circuit, measured in hertz (Hz).
frequency output: AC generator,

$$f = pN/120$$

heat loss: resistance, heat loss   *
henry: unit of inductance that induces an electromotive force of 1 V in a coil when the current is changing at the rate of 1 ampere/sec.
hertz (Hz): measure of frequency, equal to 1 cycle/sec.
horsepower:

$$hp = NT/5252$$
$$= 0.002322EI(pf)(eff)$$

ELECTRICAL — continued

impedance: opposition to current flow in an AC circuit,

$$Z = E/I$$

capacitive reactance, inductive reactance and resistance (in parallel),

$$Z = \frac{1}{[(1/R)^2 + (1/X_L - 1/X_C)^2]^{1/2}}$$

capacitive reactance, inductive reactance and resistance (in series),

$$Z = (R^2 + X^2)^{1/2}$$
$$= [R^2 + (X_L - X_C)^2]^{1/2}$$

capacitive reactance and resistance (in parallel),

$$Z = RX_C/(R^2 + X_C^2)^{1/2}$$

capacitive reactance and resistance (in series),

$$Z = (R^2 + X_C^2)^{1/2}$$

inductive reactance and resistance (in parallel),

$$Z = RX_L/(R^2 + X_L^2)^{1/2}$$

inductive reactance and resistance (in series),

$$Z = (R^2 + X_L^2)^{1/2}$$

inductance: opposition to change of current flow in a circuit. Magnetic flux surrounds a current-carrying conductor and as the current increases or decreases in the conductor, the flux increases and decreases (expands and contracts). Conversely, the magnetic flux generated around a conductor induces voltage in the conductor.

voltage induced in a DC coil,

$$E = 1 \text{ E} - 08 \ (T_C)m = Li$$

inductive reactance: reactance, inductive *
International ampere: AMPERE, INTERNATIONAL

ELECTRICAL — *continued*

International electrical units (conversion from absolute standard units):

absolute	International
ampere	= 1.000165 amperes
coulomb	= 1.000165 coulombs
farad	= 1.000495 farads
henry	= 0.999505 henry
joule	= 0.999835 joule
ohm	= 0.999505 ohm
volt	= 0.999670 volt
watt	= 0.999835 watt

joule: unit of electrical energy. It is energy required to transfer 1 coulomb of electricity between two points having a potential difference of 1 V.

$$H = 0.5E^2C = 0.5I^2L$$

Joule's law: electrical energy that converts to heat is directly proportional to the resistance of the conductor, time of current flow, and the current quantity squared.

$$H = I^2Rt_m$$

kilovar: unit of measure for reactive power in an AC circuit.

kilovolt-amperes, AC circuits:

single-phase,

$$kVA = EI/1000$$

three-phase,

$$kVA = 1.732EI/1000$$

kilowatt-hour: unit of work; total electrical energy used in a period of time,

$$kWh = kilowatts \times hours$$

ELECTRICAL — *continued*

 kilowatts:

AC single-phase circuit,

$$kW = (pf)EI/1000$$

AC three-phase circuit,

$$kW = 1.732(pf)EI/1000$$

DC circuit,

$$kW = EI/1000 = I^2R/1000 = E^2/1000/R$$

Kirchhoff's law: in a DC electrical network of conductors,

1. the algebraic sum of the currents in any one junction is zero.
2. the difference of potential between two points has a value equal to the algebraic sum of the emf's and resistance voltage drops.

locked rotor amps (LRA): the current that an electric motor with a locked rotor draws from a supply line, at a specific voltage (and frequency for AC motors).

magnetic flux: ELECTROMAGNETISM

magnetizing power: power required to produce flux for the operation of induction type equipment (as motors, transformers, etc.) measured in kilovars.

mho: unit of conductance, reciprocal of resistance unit "ohm," $G = 1/R$

motors:

efficiency,

$$eff = hp \ (output)/hp \ (input)$$

horsepower,
    AC single-phase,

$$hp = EI(eff)(pf)/746$$

    AC three-phase,

$$hp = 1.732EI(eff)(pf)/746$$

ELECTRICAL — *continued*

DC,

$$hp = EI(\text{eff})/746$$

kilowatt input,
  AC single-phase,

$$kW = EI(\text{pf})/1000$$

AC three-phase,

$$kW = 1.732EI(\text{pf})/1000$$

slip, percent,

$$S = 100(N_S - N)/N_S$$

synchronous: synchronous motor   *

neutral: power line in an AC circuit that carries current in an unbalanced circuit and no current in a balanced circuit.

ohm: unit of measure for electrical resistance to current flow, where 1 ohm allows a current of 1 ampere to pass when the electromotive force is 1 V. The standard is set as the resistance of a column of mercury at 0°C, of uniform cross-section and 106.3 cm long with a mass of 14.4521 grams.

Ohm's law: current flowing in an electrical circuit is directly proportional to the voltage and inversely proportional to the total resistance of the circuit.

$$E = IR$$
$$I = E/R$$
$$R = E/I$$

percent slip: motors, slip   *
potential: volt   *
power: rate of expending electrical energy, measured in kilovolt-amperes, kilowatts, volt-amperes, or watts.
power factor (pf): electrical term for evaluating the efficiency of an AC distribution system by measuring the ratio of true power to the apparent power. True power, measured in kilowatts, is the actual working (or usable) power. Apparent power, measured in kilovolt-amperes, is the

ELECTRICAL — *continued*

actual power flowing to the system and includes true power and the reactive (magnetizing) power (measured in kVAR) required for the operation of equipment such as motors and transformers. In the diagram shown, it is the cosine of the angle between true power and apparent power. Also, in a graph plotting current and voltage against time, it is the cosine of the angle of lag or lead between current and voltage.

power factor,

$$\text{pf} = \text{true/apparent power}$$
$$= \text{kW/kVA} = \cos B$$

single-phase,

$$\text{pf} = W/EI = W/VA = \text{kW/kVA}$$
$$= 746(\text{hp})/(EI)(\text{eff})$$

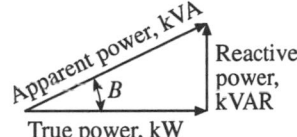

Reactive power, kVAR

True power, kW

three-phase,

$$\text{pf} = 0.5774W/EI = 577.4 \text{ kW}/EI$$
$$= 430(\text{hp})/(EI)(\text{eff})$$

power factor, percent:

$$\text{percent power factor} = 100 \times \text{power factor}$$

power factor correction: power factor may be corrected (improved or increased) by adding reactive capacitance to the circuit or system, to offset the reactive inductance in the circuit. Required reactive capac itance (measured in kVAR) is generally calculated using factors furnished by capacitor manufacturers based on the existing power factor and the corrected (or desired) power factor.

power factor correction for 100% correction,

$$\text{kVAR} = \text{kVA} \times \sin B$$

reactance, capacitive: term for the opposition to current flow in an electrical circuit due to capacitance in the circuit, measured in ohms.

$$X_c = 0.15915/fC$$

ELECTRICAL — *continued*

reactance, inductive: term for the opposition to current flow in an electrical circuit resulting from opposing voltage produced by the magnetic field of the circuit. Opposing voltage occurs as the current cycles in an AC circuit or by energizing and deenergizing a DC circuit, measured in ohms.

$$X_L = 6.283fL$$

reactance, total: algebraic sum of the capacitive reactance and the inductive reactance in an electrical circuit, measured in ohms.

$$X = X_L + (-X_C)$$

resistance: the opposition to the flow of an electrical current, measured in ohms.

$$R = E/I$$

conductor resistance, resistance of a conductor of uniform cross-section varies directly as its length and inversely as its cross-section, measured in ohms.

$$R = 1\,\mathrm{E} - 06\,rd/D^2$$
$$= 7.854\,\mathrm{E} - 07\,rd/A$$

heat loss, loss of power due to heating caused by the resistance of the conductor, measured in watts.

$$W = I^2 R$$

parallel-connected resistances, total resistance is equal to the reciprocal of the sum of the reciprocals of the individual resistances connected in parallel. two resistances in parallel,

$$R_T = R_1 R_2 / (R_1 + R_2)$$

more than two resistances in parallel,

$$1/R_T = 1/R_1 + 1/R_2 + 1/R_3 + \cdots + 1/R_n$$

series-connected resistances, total resistance is equal to the sum of the individual resistances.

$$R_T = R_1 + R_2 + R_3 + \cdots + R_n$$

ELECTRICAL — *continued*

resistivity coefficient (*a*) for temperature correction for various materials (units per °F):

aluminum: 0.00239	lithium: 0.0026
antimony: 0.002	magnesium: 0.002
barium: 0.0018	manganin: 1.39 E−05
beryllium: 0.0022	mercury: 4.94 E−04
bismuth: 0.0022	molybdenum: 0.0026
brass: 0.0011	monel: 0.0013
bronze: 0.0011	nichrome: 5.6 E−05
cadmium: 0.0023	nickel: 0.003
calcium: 0.002	niobium: 0.0022
carbon: −0.0005	osmium: 0.0018
chromium: 0.0016	palladium: 0.0021
cobalt: 0.00366	platinum: 0.0022
constantan: 1.11 E−06	potassium: 0.0031
copper, annealed: 0.00217	rhodium: 0.00239
hard drawn: 0.00211	silver: 0.0021
pure: 0.00218	sodium: 0.003
german silver: 1.4 E−04	steel: 0.0024
gold: 0.00189	tantalum: 0.0017
graphite: 5 E−04	thallium: 0.0022
inconel: 5.5 E−07	thorium: 0.00117
indium: 0.0028	tin: 0.0023
invar: 6.7 E−04	titanium: 0.0033
iron: cast: 0.0036	tungsten: 0.0026
wrought: 0.0022	zinc: 0.0023
lead: 0.0022	zirconium: 0.0024

resistivity (*r*) of a material: the measured resistance to electrical current of a material of a specific cross-section and length at a known temperature, and is the reciprocal of conductivity, measured in ohm-mil/ft.

$$r = 1/g$$

at other temperatures,

$$r_2 = r_1[1 + a(t_2 - t_1)]$$

ELECTRICAL — *continued*

resistivity (*r*) of various materials (ohm-mil/ft):

aluminum, annealed: 16.73	indium: 49.4
hard drawn: 17.00	invar (64Fe, 36Ni): 519
pure: 16.21	iridium: 37.9
amber: 3 E+23	iron, cast: 415
antimony: 253	pure: 57.6
asbestos paper: 9.6 E+17	wrought: 86.4
bakelite: 3 E+22	ivory: 1.2 E+15
barium: 57.6	lead: 133
beeswax: 3.6 E+21	lithium: 52
beryllium: 111	magnesium: 26.6
bismuth: 741	manganese: 1152
brass, annealled: 41.5	manganin: 265
yellow: 37	marble: 3 E+17
bronze: 23	mercury: 576
bronze: phosphorus: 27.3	methyl alcohol: 8.4 E+11
cadmium: 41.5	mica: 6 E+23
calcium: 28	molybdenum: 33.46
carbon: 2.1 E+04	monel: 258
celluloid: 1.2 E+17	nichrome: 610
cerium: 471	nickel: 47
cesium: 114	niobium: 74
chromium: 16.7	olive oil: 3 E+19
cobalt: 57.6	osmium: 57.6
columbium: 72.2	palladium: 64.8
constantan: 296	paper: 3 E+21
copper, annealed: 10.37	paraffin: 1.5 E+25
hard drawn: 10.8	paraffin oil: 6 E+23
pure: 10.37	petroleum oil: 1.2 E+23
ethyl alcohol: 1.8 E+12	platinum: 59.3
germanium: 280	porcelain: 1.8 E+21
german silver: 247	potassium: 41.5
glass: 1.5 E+23	quartz: 1.5 E+25
gold: 14.6	rhodium: 33.5
graphite: 5186	rosin: 1.7 E+23
hafnium: 230	rubber, hard: 3 E+24
ice: 4.3 E+15	ruthenium: 45.7
inconel: 610	selenium: 72

ELECTRICAL — *continued*

shellac: 6 E+22	tantalum: 74.6
silica, fused: 3 E+25	tellurium: 1.2 E+06
silicon: 6 E+05	thallium: 105.8
silver: 9.88	thorium: 108
slate: 3 E+16	tin: 69
sodium: 31.6	titanium: 288
steel, carbon: 138	tungsten: 33.5
silicon: 305	uranium: 175.8
soft: 94.3	vanadium: 154.8
stainless: 432	water, distilled: 3 E+12
structural: 86	wood, paraffined: 1.2 E+20
strontium: 150	zinc: 34.6
sulfur: 4.8 E+22	zirconium: 253

single phase: an AC electrical circuit energized by a single alternating voltage, alternating from a maximum positive voltage to a minimum negative voltage, at a specific frequency and following a sine curve.

slip: difference between the speed of the rotating field of a motor and its rotor speed.

slip, percent: motors, slip   *

steel wire gauges, resistances and weights: STEEL WIRE GAUGES

synchronous motor: an AC electric motor with a separately excited DC field to maintain a constant speed of the rotor, often used in AC circuits for power factor correction.

synchronous speed: rotating speed of the rotor of a motor.

$$N_S = 120f/p$$

three phase: an AC electrical circuit energized by a combination of three alternating voltages, each alternating from a maximum positive voltage to a minimum negative voltage, alternating at a specific frequency, in sine-wave paths spaced at 120°.

delta-connected, 3-phase, 3-wire,

ELECTRICAL — *continued*

wye-connected, 3-phase, 3-wire,

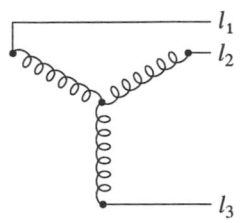

wye-connected, 3-phase, 4-wire,

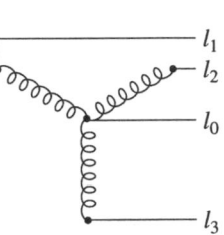

total reactance: reactance, total   *

transformers: an electrical device for increasing (or decreasing) electrical voltage. It consists of two separate coil windings, the primary (input) and the secondary (output). Energy is transferred from the primary coil to the secondary coil through induction produced by the magnetic field of the primary coil. Current flowing in the primary coil winding creates a magnetic flux in the common iron core shared by the primary and secondary windings and thereby induces a current flow in the secondary winding. The secondary voltage is determined by the ratio of number of turns of the primary coil to the number of turns in the secondary coil.

transformer, autotransformer: autotransformer   *
transformer, single-phase,
    capacity,

$$\text{kVA} = IE/1000$$

    current, primary,

$$I_P = 1000 \ \text{kVA}/E_P$$

ELECTRICAL — *continued*

current, secondary,

$$I_S = 1000 \text{ kVA}/E_S$$

transformer, three-phase,
capacity,

$$\text{kVA} = 1.732IE/1000$$

current, primary,

$$I_P = 577 \text{ kVA}/E_P$$

current, secondary,

$$I_S = 577 \text{ kVA}/E_S$$

transformer, two-phase, 4-wire,
capacity,

$$\text{kVA} = IE/500$$

transformer current by turns ratio,
primary,

$$I_P = I_S\, T_S/T_P$$

secondary,

$$I_S = I_P\, T_P/T_S$$

transformer voltage,
secondary,

$$E_S = E_P\, T_S/T_P$$

true power: power factor   ∗

two phase: an electrical circuit energized by a combination of two alternating voltages, each alternating from a positive maximum voltage to a negative, in sine-wave paths and spaced at 90°, alternating at a specific frequency (as 60 Hz).

volt: electromotive force, emf, potential difference between two points. One volt produces a current of 1 ampere through resistance of 1 ohm.

ELECTRICAL — *continued*

voltage:

AC circuits,
impedance only,

$$E = I/Z$$

inductive reactance and resistance,

$$E = I(R^2 + X_L^2)^{1/2}$$

resistance only,

$$E = IR$$

single-phase, 2-wire circuit,

$$E = kW/I/pf$$

three-phase, 3-wire, delta-connected, balanced load,

$$E = 0.577IR$$
$$= 0.577\,kW/I/pf$$

three-phase, 3-wire, wye-connected, balanced load,

$$E \text{ (phase to phase)} = 1.732E \text{ (phase to common)}$$

three-phase, 4-wire, wye-connected, balanced load,

$$E \text{ (phase to phase)} = 1.732E \text{ (phase to neutral)}$$

DC circuits,

$$E = IR = VA/I$$

voltage drop: loss of voltage in an electrical circuit across terminals of the resistance.

voltage drop, due to resistance of circuit feed wire:

AC circuit, single-phase,

$$VD = 0.002\,IZd_1$$

AC circuit, three-phase, phase to neutral,

$$VD = 0.002\,IZd_1$$

ELECTRICAL — *continued*

AC circuit, three-phase, phase to phase,

$$VD = 0.001732\, IZd_1$$

DC circuit,

$$VD = 0.002\, IRd_1$$

volt-ampere: power unit in an AC circuit,

single-phase,

$$VA = EI$$

three-phase,

$$VA = 1.732EI$$

watt: power unit, energy of 1 W is produced between two points by 1 ampere of current flowing under a potential of 1 volt.

AC circuit, single-phase,

$$W = EI(\text{pf}) = VA(\text{pf}) = I^2R$$

AC circuit, three-phase,

$$W = 1.732EI(\text{pf})$$
$$= 1.732VA(\text{pf})$$

DC circuit,

$$W = EI = E^2/R = I^2R$$

watt-hour: electrical energy supplied to a circuit for a time period.

$$\text{Wh} = \text{watt} \times \text{hour}$$

electric heat: HEAT TRANSFER

ELECTROLYSIS: process where an electric voltage is connected across positive and negative electrodes submersed in an electrolyte. A chemical reaction occurs at the electrodes causing specific substances to be deposited on the electrodes or a gas is released at the electrodes.

ELECTROLYSIS — *continued*

electrolyte: chemical solution of a substance dissolved in a liquid and forming ions in the solution that conduct an electrical current through the solution.

electrolytic cell: container of electrolyte solution and positive and negative electrodes.

electromotive series: electromotive force series, galvanic series, listing of the electrode potential of elements and other materials according to decreasing tendency to release electrons, with hydrogen as zero.

electromotive series of various materials (V)

cesium: +3.02	tin: +0.14
lithium: +3.02	lead: +0.13
rubidium: +2.99	HYDROGEN: 0.00
potassium: +2.92	stainless steel: −0.09
barium: +2.90	antimony: −0.10
strontium: +2.89	bismuth: −0.226
calcium: +2.87	silicon: −0.26
sodium: +2.71	arsenic: −0.30
magnesium: +2.40	oxygen: −0.397
lanthanum: +2.37	polonium: −0.40
beryllium: +1.70	brass: −0.47
aluminum: +1.67	bronze: −0.47
uranium: +1.4	copper: −0.47
manganese: +1.05	monel: −0.47
tellurium: +0.827	iodine: −0.53
zinc: +0.76	lead: −0.80
chromium: +0.71	silver: −0.80
sulfur: +0.51	palladium: −0.83
gallium + 0.50	mercury: −0.85
steel: +0.45	bromine: −1.06
iron: +0.44	platinum: −1.20
cadmium: +0.40	water: −1.23
indium: +0.336	chlorine: −1.36
thallium: +0.33	gold: −1.42
cobalt: +0.28	fluorine: −3.03
nickel: +0.25	

faraday: quantity of electricity equal to 96,490 coulombs.

ELECTROLYSIS — continued

Faraday's law of electrolysis: the number of gram-equivalent weights of a substance deposited or liberated at an electrode is equal to the number of faradays passed through an electrolyte, as 1 faraday (96,490 coulombs) produces 1 gram-equivalent weight of a substance.

galvanic cell: an electrolytic cell consisting of two dissimilar electrodes (anode and cathode) in an electrolytic solution and capable of producing electrical energy by chemical reaction.

galvanic series: electromotive series *

ELECTROMAGNETISM: where an electric current flowing through an electrical current-carrying conductor made of soft iron or steel produces a magnetic field around the conductor, where

a = cross-sectional area of wire or cylinder, cm^2
A = cross-sectional area of induction lines, cm^2
B = flux density, gauss
c = hysteresis coefficient of the material
d = distance, cm
e = voltage induced, volts
E = energy, joules
emf = electromotive force
f = magnetic flux, maxwells
F = force, dyn
h = hysteresis loss, $erg/cm^3/cycle$
H = magnetic intensity, oersted or gilbert/cm
I = current, ampere
l_c = length of conductor or cylinder, cm
l_F = length of flux, cm
L = inductance, henrys
M = magnetomotive force, gilberts
n = number of turns/cm of length
N = number of turns
P = permeance, maxwells/gilbert
r = radius of circular turn of wire, cm
R = reluctance, gilberts/maxwell
u = permeability (air = 1)
V = velocity, cm/sec

ELECTROMAGNETISM — *continued*

ampere-turn: the product of the current and number of turns in the current-carrying winding in an electromagnetic circuit, as in a solenoid.

coefficient of hysteresis: hysteresis coefficient   *

eddy current: current induced in an object by a magnetic field.

eddy current losses: electrical energy losses in the electromagnetic process.

energy: energy produced by self-induction, $E = 0.5\,I^2L$

Faraday's law: induced emf in an electromagnetic circuit is numerically proportional to the rate of change of magnetic flux.

flux density: magnetic flux per unit area, $B = f/A$

force: pressure due to magnetic field.

conductor in a magnetic field,

$$F = 0.1BIlc$$

between two parallel current-carrying conductors,

$$F = 0.02\,I_1 I_2/d$$

gauss: metric unit for density or magnetizing flux in an electromagnetic system.

gauss/oersted: metric unit for permeability.

gilbert: metric unit for magnetomotive force, or magnetic potential in the electromagnetic system.

gilbert/maxwell: metric unit for reluctance.

$$R = 7.958\,\mathrm{E} + 07 \text{ ampere-turns/weber}$$

hysteresis coefficient (*c*) for various materials

iron, cast: 0.013	steel, low-carbon sheet: 0.003
iron, sheet: 0.004	steel, 50% nickel: 0.00015
steel, cast: 0.005	steel, silicon: 0.0009
steel, forged: 0.020	steel, silicon sheet: 0.001

hysteresis loop: electrical term for the magnetization cycle, indicating flux density variation and lag as the magnetizing force changes from positive to negative and back from negative to positive, as in an alternating current.

ELECTROMAGNETISM — *continued*

hysteresis loss: energy loss in the magnetization process of a material due to friction of molecules as flux is changed. Loss varies with different materials.

$$h = c(B)^{1.6}$$

inductance:

coil,

$$L = 1 \text{ E} - 08 \ FN/I$$

solenoid,

$$L = 12.6 \text{ E} - 09 \ n^2 ua$$

Lenz's law: in an electromagnetic induction circuit, the direction of the current of the induced emf opposes the motion producing it.
line: line of induction, maxwell
lines per unit area: lines of induction per unit area.
magnetic flux: total lines of induction produced by a current-carrying conductor.

$$f = M/R = BA$$

magnetic intensity (*H*): strength of magnetic field force.

center of several circular turns of current-carrying wire,

$$H = 0.628 \ NI/r$$

center of single circular turn of wire,

$$H = 0.628 \ I/r$$

point, near center along an axis of an axially wound coil,

$$H = 1.26nI$$

point, perpendicular from a current-carrying wire, where *d* is greater than the radius of wire and insignificant to the length of wire,

$$H = 0.21/d$$

ELECTROMAGNETISM — *continued*

magnetizing power: ELECTRICAL

magnetomotive force: energy that produces magnetic flux proportional to the ampere-turns of the circuit.

$$M = 1.257NI$$

permeability: ratio of magnetic flux of a material to magnetic flux of air in the same magnetic field.

$$u = B/H$$

permeance: conductivity permitting flow of magnetic flux, the reciprocal of reluctance.

$$P = I/R$$

reluctance: resistance to flow of magnetic flux, and is directly proportional to the length of the material (parallel to magnetic lines) and inversely proportional to the cross-sectional area of the magnetic flux and permeability of the material.

$$R = l_F/Au$$

voltage, induced: voltage induced by movement through a magnetic field.

single conductor,

$$e = 1 \text{ E}-08 \; Bl_C \, V$$

coil, several turns of a conductor,

$$e = 1 \text{ E}-08 \; Bl_C \, NV$$

---

electromotive force: ELECTRICAL

electromotive series: ELECTROLYSIS

electromotive series of various materials: ELECTROLYSIS

electron: negative charged particle orbiting the nucleus of an atom.

electron charge =

1.603 E − 19 C

4.802 E − 10 statcoulomb

electron mass $= 9.107 \text{ E} - 28$ g
electron volt $=$

1.602 E $- 12$ erg
1.602 E $- 19$ J

ELECTROSTATICS: form of static electricity where an object is charged with a negative (or positive) current; it retains the charge until it is leaked out or discharged by contact with a second object of opposite polarity. Opposite charges attract and like charges repel.

electrostatics formulas: where [units shown are cgs and (mks)]

A = surface area, cm^2 (m^2)
c = capacitance per cylinder length, microfarads/cm
C = capacitance, microfarads (farads)
C_1, C_2 = individual capacitances, microfarads (farads)
C_T = total capacitance, microfarads (farads)
d = diameter of cylinder, cm
d_1 = outside diameter of inner cylinder, cm
d_2 = inside diameter of outer cylinder, cm
E = potential, statvolts (volts)
$E E_1, E_2, E_3,$ = potential across individual
   capacitors, statvolts (volts)
E_T = total potential, statvolts (volts)
f = field intensity, dyn/statcoulomb or
   statvolts/cm (Newtons/coulomb or volts/meter)
F = force, dyn (Newtons)
I = current, ampere
k = dielectric constant of medium
L = dielectric flux, number of lines per unit area
P = permittivity of dielectric material,
   statcoulomb^2/dyn-cm^2 (coulomb^2/Newton-meter^2)
Q = quantity of charge, statcoulombs (coulombs)
Q_1, Q_2 = charges of poles 1 and 2 statcoulombs (coulombs)
Q_A = charge per unit area, statcouluombs/cm^2 (coulomb/meter^2)
Q_T = total charge, statcouluombs (coulombs)
S = space or dielectric thickness, cm (m)
t = time, sec
V = voltage gradient, statvolts/cm (volts/meter)
W = energy stored, erg (Joules)

ELECTROSTATICS — *continued*

capacitance: ELECTRICAL
capacitance formulas:

capacitor,

$$\text{cgs units, } C = Q/(9 \text{ E} + 05 \text{ } E)$$
$$\text{mks units, } C = Q/E$$

capacitors in parallel,

$$C_T = C_1 + C_2 + C_3 + \cdots + C_n$$

capacitors in series,

$$1/C_T = 1/C_1 + 1/C_2 + 1/C_3 + \cdots + 1/C_n$$

cylindrical capacitor, two coaxial cylinders with dielectric material between the cylinders,

$$c = 6.13 \text{ E} - 07 \text{ } k/\log (d_2/d_1)$$

electric cables: ELECTRICAL
parallel cylinders, two, same diameter,

$$c = 3.063 \text{ E} - 07 \text{ } k/\log (2S/d)$$

parallel plate condenser,

$$\text{cgs units, } C = 8.842 \text{ E} - 08 \text{ } kA/S$$
$$\text{mks units, } C = 8.842 \text{ E} - 12 \text{ } kA/S$$

capacitivity: dielectric constant   *
capacitor: ELECTRICAL
capacitor charge formulas:

capacitor,

$$\text{cgs units, } Q = 9 \text{ E} + 05 \text{ } CE$$
$$\text{mks units, } Q = CE$$

capacitors in parallel,

$$Q_T = Q_1 + Q_2 + Q_3 + \cdots + Q_n$$

capacitors in series,

$$Q_T = Q_1 = Q_2 = Q_3 = \cdots = Q_n$$

ELECTROSTATICS — *continued*

charge per unit area,

$$Q_A = Q/A$$

time required to charge capacitor,

cgs units, $t = 3.333 \text{ E} - 10 \; Q/I = 3 \text{ E} - 04 \; CE/I$
mks unit, $t = Q/I = CE/I$

coulomb: quantity unit of electrical charge flowed, through a section of a conductor, in 1 sec by a current of 1 ampere, equals 1 ampere-seconds.

Coulomb's law: the force between two poles is directly proportional to the charges of the poles and inversely proportional to the square of the distance between the poles.

cgs units,

$$F = Q_1 Q_2 / S^2$$

mks units,

$$F = 9 \text{ E} + 09 \; Q_1 Q_2 / S^2$$

daraf: reciprocal of capacitance.
dielectric: insulating material between two electrostatic charges.
dielectric coefficient: dielectric constant *
dielectric constant: capacitivity, dielectric coefficient, ratio of insulating quality (P) of a medium (material) compared to vacuum,

cgs units, $k = P/0.0796 = 12.57P$
mks units, $k = P/8.842 \text{ E} - 12 = 1.131 \text{ E} + 11 \; P$

dielectric constant (k) for various materials:

acetic acid: 6.2
acetone: 20.7
air: 1.006
asbestos paper: 2.7
asphalt: 2.7
bakelite: 4.5–5.5
benzene: 2

carbon dioxide: 1.001
carbon monoxide: 1.001
carbon tetrachloride: 2.2
castor oil: 4.7
cellophane: 8
celluloid: 13.3
cellulose acetate: 5

ELECTROSTATICS — *continued*

ebonite: 3
ether: 4.3
ethyl alcohol: 24.3
ethylene glycol: 37.7
fiber: 2.5–5.0
glass: 5.4–9.1
glyerine: 56
hydrogen: 1.003
hydrogen bromide: 6
hydrogen iodide: 3
ice: 86
linseed oil: 3.3
marble: 8
masonite: 3
methyl alcohol: 32
mica: 2.5–6.6
oil: 2.2–4.7
olive oil: 3
paper: 1.7–3.8

paraffin: 1.9–2.3
petroleum oil: 2
polyethylene: 2.4
porcelain: 5
quartz: 4.7–5.1
rosin: 2.5
rubber, hard: 1.9–3.5
selenium: 6.1–7.4
shellac: 3.0–3.7
silica, fused: 3.5
slate: 6.6–7.4
steam: 1.007
sulfur: 2.9–3.2
turpentine: 2
varnished cambric: 4–6
vacuum: 1.000
water, distilled: 80
wood: 2.5–8.0
wood, paraffined: 4

dielectric flux: number of lines of a charged body,

$$\text{cgs units, } L = 12.57Q$$
$$\text{mks units, } L = 3.77 \text{ E}+10 \ Q$$

dielectric intensity: field intensity *

dielectric strength: resistance to electrostatic breakdown. It is the limit value where the dielectric becomes a conductor (which varies with humidity, pressure, and temperature), rated in statvolts/cm thickness.

dielectric strength values for various materials (statvolts/cm; multiply by 0.761 for volts/mil):

air: 100
asphalt: 130–525
bakelite: 200–650
ebonite: 1000–3700
fiber: 66
glass: 1000–5000
marble: 80
masonite: 150–250
mica: 1000–7300

oil: 130–650
paper: 325–790
paraffin: 260–1500
porcelain: 130–330
rubber: 525–1600
varnished cambric: 1050
water: 500
wood: 33–100

ELECTROSTATICS — *continued*

energy stored in a capacitor (condenser):

mks units, $W = 0.5QE = 0.5E^2C = 0.5Q^2/C$

farad: ELECTRICAL

field intensity: dielectric intensity, force per unit charge on a point distance (S) from charged body,

$$f = Q/S^2P = F/Q$$

forces:

charged body in a dielectric field,

$$F = Qf$$

two charged bodies,

$$F = Q_1Q_2/S^2P$$

permittivity:

vacuum, $k = 1$

cgs units, $P = 0.0796$
mks units, $P = 8.84 \text{ E} - 12$

other media,

cgs units, $P = 0.0796k$
mks units, $P = 8.84 \text{ E} - 12 \, k$

potential: voltage ✱

statcoulomb: electrostatic unit for quantity of charge that repels an equal charge at a distance of 1 cm apart, with a force of 1 dyn.

statvolt = 1 erg/statcoulomb

volt: work done in moving a charge between two points,

$$V = 1 \text{ joule/coulomb}$$

voltage: potential,

across capacitors in parallel,

$$E_T = E_1 = E_2 = E_3 = \cdots = E_n$$

ELECTROSTATICS — *continued*

across capacitors in series,

$$E_T = E_1 + E_2 + E_3 + \cdots + E_n$$

between point and a charged body,

$$E = Q/P/S$$

two parallel plates,

$$E = 5S(FPA)^{1/2}$$

voltage gradient: voltage across a dielectric per unit thickness of the dielectric,

$$V = Q_A/P = E/S$$

elements (as listed in the Periodic Table)

Element	Atomic No.	Symbol	Atomic Weight
Actinum	89	Ac	227
Aluminum	13	Al	26.97
Americium	95	Am	[243]
Antimony	51	Sb	121.76
Argon	18	A	39.944
Arsenic	33	As	74.91
Astatine	85	At	[211]
Barium	56	Ba	137.36
Berkelium	97	Bk	[245]
Beryllium	4	Be	9.02
Bismuth	83	Bi	209
Boron	5	B	10.82
Bromine	35	Br	79.916
Cadmium	48	Cd	112.41
Calcium	20	Ca	40.08
Californium	98	Cf	[246]

elements — *continued*

Element	Atomic No.	Symbol	Atomic Weight
Carbon	6	C	12.01
Cerium	58	Ce	140.13
Cesium	55	Cs	132.91
Chlorine	17	Cl	35.457
Chromium	24	Cr	52.01
Cobalt	27	Co	58.94
Columbium	41	Cb	92.91
Copper	29	Cu	63.54
Curium	96	Cm	[243]
Dysprosium	66	Dy	162.46
Erbium	68	Er	167.2
Europium	63	Eu	152
Fluorine	9	F	19
Francium	87	Fr	[223]
Gadolinium	64	Gd	156.9
Gallium	31	Ga	69.72
Germanium	32	Ge	72.60
Gold	79	Au	197.2
Hafnium	72	Hf	178.6
Helium	2	He	4.003
Holmium	67	Ho	164.94
Hydrogen	1	H	1.008
Indium	49	In	114.76
Iodine	53	I	126.92
Iridium	77	Ir	193.1
Iron	26	Fe	55.85
Krypton	36	Kr	83.7
Lanthanum	57	La	138.92
Lead	82	Pb	207.21
Lithium	3	Li	6.94
Lutecium	71	Lu	174.99
Magnesium	12	Mg	24.32

elements — *continued*

Element	Atomic No.	Symbol	Atomic Weight
Manganese	25	Mn	54.93
Mercury	80	Hg	200.61
Molybdenum	42	Mo	95.95
Neodymium	60	Nd	144.27
Neon	10	Ne	20.183
Neptunium	93	Np	[237]
Nickel	28	Ni	58.69
Niobium	41	Nb	92.91
Nitrogen	7	N	14.008
Osmium	76	Os	190.2
Oxygen	8	O	16.00
Palladium	46	Pd	106.7
Phosphorus	15	P	30.98
Platinum	78	Pt	195.23
Plutonium	94	Pu	[242]
Polonium	84	Po	210
Potassium	19	K	39.096
Praseodymium	59	Pr	140.92
Promethium	61	Pm	[145]
Protactinium	91	Pa	231
Radium	88	Ra	226.05
Radon	86	Rn	222
Rhenium	75	Re	186.31
Rhodium	45	Rh	102.91
Rubidium	37	Rb	85.48
Ruthenium	44	Ru	101.7
Samarium	62	Sm	150.43
Scandium	21	Sc	44.96
Selenium	34	Se	78.96
Silicon	14	Si	28.06
Silver	47	Ag	107.88
Sodium	11	Na	22.997

elements — *continued*

Element	Atomic No.	Symbol	Atomic Weight
Strontium	38	Sr	87.63
Sulfur	16	S	32.066
Tantalum	73	Ta	180.88
Technetium	43	Tc	[99]
Tellurium	52	Te	127.61
Terbium	65	Tb	159.2
Thallium	81	Tl	204.39
Thorium	90	Th	232.12
Thulium	69	Tm	169.4
Tin	50	Sn	118.7
Titanium	22	Ti	47.9
Tungsten	74	W	183.92
Uranium	92	U	238.07
Vanadium	23	V	50.95
Xenon	54	Xe	131.3
Ytterbium	70	Yb	173.04
Yttrium	39	Y	88.92
Zinc	30	Zn	65.38
Zirconium	40	Zr	91.22

[ ] indicates an isotope.

ELEVATOR FORMULAS: where

a = angle of travel, degrees
eff = efficiency, decimal
hp = horsepower
V = speed, ft/min
W = weight (unbalanced load), lb

inclined elevator, horsepower,

$$\text{hp} = \frac{VW \sin a}{33,000 \times \text{eff}}$$

ELEVATOR FORMULAS — *continued*

 vertical elevator, horsepower,

$$\text{hp} = \frac{VW}{33,000 \times \text{eff}}$$

ell: English measure of cloth length
ell =

114.3 cm   45 in.
3.75 ft

ELLIPSE: where

a = length, major axis, in.
A = area, in.^{2}
b = width, minor axis, in.
$FP + F'P$ = constant

area,

$$A = 0.7854ab$$

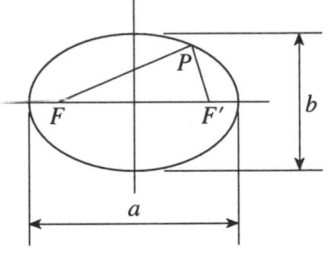

ELLIPSE SECTION FORMULAS: where

a, b, c, d, x, y = dimensions, in.
A = area, in.^{2}
cg = center of gravity
I_{AA}, I_{BB} = moment of inertia, in.^{4}
I_P = polar moment of inertia, in.^{4}
k_{AA}, k_{BB} = radius of gyration, in.
Z_{AA}, Z_{BB} = section modulus, in.^{3}

ellipse, hollow (elliptical ring):

area, $A = 0.7854(ab - cd)$
center of gravity,

$$x = 0.5a$$
$$y = 0.5b$$

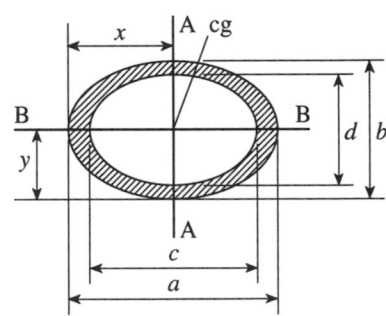

ELLIPSE SECTION FORMULAS — *continued*

moment of inertia,

$$I_{AA} = 0.0491(a^3b - c^3d)$$
$$I_{BB} = 0.0491(b^3a - d^3c)$$

radius of gyration,

$$k_{AA} = \frac{0.25(a^3b - c^3d)^{1/2}}{(ab - cd)^{1/2}}$$

$$k_{BB} = \frac{0.25(b^3a - d^3c)^{1/2}}{(ab - cd)^{1/2}}$$

section modulus,

$$Z_{AA} = 0.0982(a^3b - c^3d)/a$$
$$Z_{BB} = 0.0982(b^3a - d^3c)/b$$

ellipse, solid:

area,

$$A = 0.7854ab$$

center of gravity,

$$x = 0.5a$$
$$y = 0.5b$$

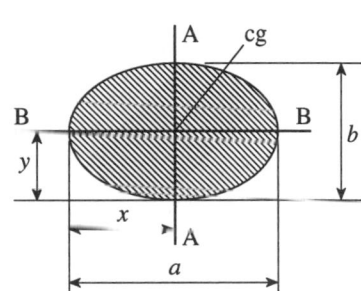

moment of inertia,

$$I_{AA} = 0.0491a^3b$$
$$I_{BB} = 0.0491b^3a$$

moment of inertia, polar,

$$I_P = 0.0491ab(a^2 + b^2)$$

radius of gyration,

$$k_{AA} = 0.25a$$
$$k_{BB} = 0.25b$$

ELLIPSE SECTION FORMULAS — *continued*

section modulus,

$$Z_{AA} = 0.0982a^2b$$
$$Z_{BB} = 0.0982b^2a$$

ELLIPSOID (THREE-DIMENSIONAL SOLID ELLIPSE): where

a = major axis, ft
b = minor axis, ft
c = depth axis, ft
I_{AA}, I_{BB}, I_{CC} = moment of inertia, slug-ft^2
k_{AA}, k_{BB}, k_{CC} = radius of gyration, ft
V = volume, ft^3
W = weight, lb

moment of inertia,

$$I_{AA} = 0.00155W(b^2 + c^2)$$
$$I_{BB} = 0.00155W(a^2 + c^2)$$
$$I_{CC} = 0.00155W(a^2 + b^2)$$

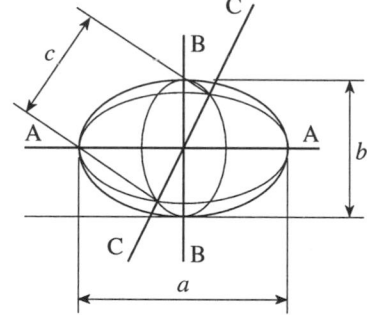

radius of gyration,

$$k_{AA} = 0.2236(b^2 + c^2)^{1/2}$$
$$k_{BB} = 0.2236(a^2 + c^2)^{1/2}$$
$$k_{CC} = 0.2236(a^2 + b^2)^{1/2}$$

volume,

$$V = 0.5236abc$$

emissivity: HEAT TRANSFER
engineer's chain: CHAIN (RAMDEN)
engineer's measure: LINK (RAMDEN)
enthalpy: THERMODYNAMICS
entropy change: THERMODYNAMICS
equilateral triangle: TRIANGLES
equivalent weight: GRAM EQUIVALENT WEIGHT
Er: erbium (element)

erg: dyne-centimeter; metric energy unit for work done by a force of 1 dyn acting through a distance of 1 cm.

erg =

9.486 E − 11 Btu	3.775 E − 14 hp-h (metric)
9.869 E − 07 cm³-atm	8.8507 E − 07 in.-lb
1 dyn-cm	1 E 07 J
6.2415 E + 11 eV	2.39 E − 11 kg-cal
3.4853 E − 11 ft³-atm	1.0197 E − 08 kg-m
7.3756 E − 08 ft-lb	2.778 E − 14 kWh
2.373 E − 06 ft-pdl	9.87 E − 10 l-atm
2.39 E − 08 g-cal	1 E − 07 N-m
0.0010197 g-cm	2.778 E − 11 Wh
3.725 E − 14 hp-h	1 E − 07 W-s

erg/minute = 4.4254 E − 06 ft-lb/h
erg/second: metric unit for power
erg/second =

3.414 E − 07 Btu/h	0.0010197 g-cm/s
5.69 E − 09 Btu/min	1.341 E − 10 hp
9.48 E − 11 Btu/s	1.0194 E − 11 hp (boiler)
1 dyn-cm/s	1 E − 07 J/s
4.425 E − 06 ft-lb/min	1.434 E − 09 kg-cal/min
7.378 E − 08 ft-lb/s	2.39 E − 11 kg-cal/s
8.604 E − 05 g-cal/h	1.0197 E − 08 kg-m/s
1.434 E − 06 g-cal/min	1 E − 10 kW
2.39 E − 08 g-cal/s	1 E − 07 W

erg/square centimeter: DYNE/CM
erg/square millimeter =

100 dyn/cm	100 erg/cm²

ethanol: ethyl alcohol, grain alcohol
ether: diethyl ether
ethyl alcohol: ethanol, grain alcohol
Eu: europium (element)
Euler's formula: COLUMN LOAD LIMITS

F

F: fluorine (element)

fahrenheit degrees: °F

FAN FORMULAS: where

A = area of discharge, in.^{2}
bhp = brake horsepower
eff = efficiency, decimal
h = pressure rise across fan, in. (water)
h_S = static pressure, in. (water)
h_T = total pressure, in. (water)
h_V = velocity pressure, in. (water)
hp = horsepower, air horsepower
p = gauge pressure, psig
Q = flow capacity, ft^3/min
Q_D = discharge capacity, ft^3/min
V = flow velocity, ft/min
w = specific weight, lb/ft^3
W = weight flow, lb/min

air horsepower: horsepower output of fan,

$$hp = Qh_T/6350$$
$$= pQ/229$$
$$= 2.5\,E-05\,Wh$$

brake horsepower: horsepower input required at the fan driveshaft,

$$bhp = hp/eff$$

FAN FORMULAS — *continued*

flow capacity, discharge:

$$Q_D = AV/144$$

flow by weight:

$$W = wQ_D$$

horsepower: air horsepower   *

static pressure: pressure of a gas in a pipe (or duct), exerted in all directions, measured perpendicular to the flow of the material, and is basically the pressure loss due to friction and other resistances.

total pressure: the sum of static and velocity pressures created by the fan,

$$h_T = h_S + h_V$$

velocity:

$$V = 1098(h_V/w)^{1/2}$$
$$= 144Q_D/A$$

velocity pressure: pressure in the direction of flow of the material, the difference between total pressure measured and static pressure measured,

$$h_V = 8.3\,\text{E} - 07\ V^2 w$$
$$= h_T - h_S$$

FAN LAWS: where

$$D_1, D_2 = \text{fan diameter, initial and final, ft}$$
$$\text{hp}_1, \text{hp}_2 = \text{fan horsepower, initial and final}$$
$$N_1, N_2 = \text{fan speed, initial and final, rev/min}$$
$$h_{S1}, h_{S2} = \text{static pressure, initial and final, in. (water)}$$
$$Q_1, Q_2 = \text{fan flow capacity, initial and final, ft}^3/\text{min}$$
$$w_1, w_2 = \text{specific weight of the gas, initial and final, lb/ft}^3$$

flow capacity varies directly as the fan speed ratio,

$$Q_2 = Q_1 N_2/N_1$$
$$= Q_1 D_2^3/D_1^3$$

FAN LAWS — *continued*

horsepower varies directly as the cube of the fan speed ratio,

$$hp_2 = hp_1 N_2^3/N_1^3$$
$$= hp_1 w_2/w_1$$
$$= hp_1 D_2^5/D_1^5$$

static pressure varies directly as the square of the fan speed ratio,

$$h_{S2} = h_{S1} N_2^2/N_1^2$$
$$= h_{S1} w_2/w_1$$
$$= h_{S1} D_2^2/D_1^2$$

farad: ELECTRICAL
farad =

1 E−09 abfarad	1 E+12 pF
1.00052 International farads	8.9876 E+11 statfarads
1 E+06 μF	

faraday: ELECTROLYSIS
faraday =

9648 abcoulombs	9.649 E+04 C
26.8 A-h	6.06 E+23 electron charges
9.649 E+04 A-s	2.8924 E+14 statcoulombs

faraday/second =

9648 abamperes
9.648 E+04 A

Faraday's law: ELECTROMAGNETISM
Faraday's law of electrolysis: ELECTROLYSIS
fathom =

0.008333 cable length	72 in.
182.88 cm	3.2916 E−04 league (naut)
6 ft	3.7879 E−04 league (stat)

fathom — *continued*

1.8288 m	0.0011364 mi (stat)
9.8747 E − 04 mi (naut)	2 yd

fatigue failure: STRENGTH OF MATERIALS
Fe: iron (element)
femto: prefix, equals 1 E − 15
flashpoint: the lowest temperature whereby a flammable liquid vaporizes and ignites when exposed to an open flame
flashpoint of various materials (°F)

acetic acid: 108
acetone: 0
asphalt: 400
benzene, benzol: 12
benzine: 0
camphor oil: 117
castor oil: 450
coal tar pitch: 405
coconut oil: 420
corn oil: 490
cotton seed oil: 486
creosote oil: 165
diesel fuel: 100
dry cleaning solvent: 105
ethyl acetate: 26
ethyl alcohol: 55
ethylene glycol: 230
fish oil: 420
formic acid: 156
fuel oil no. 1: 115–180
    no. 2: 125–230
    no. 3: 125–230
    no. 4: 154–240
    no. 5: 130–310
    no. 6: 150–430

gasoline: −45
glycerine: 320
heptane: 25
isopropyl alcohol: 61
kerosene: 110
lanolin: 460
lard oil: 390
linseed oil: 400
methyl alcohol: 52
methyl ethyl ketone: 30
mineral oil: 300
mineral spirits: 100
naphtha: 105
naphtha VMP: 30
octane: 56
oleo oil: 450
olive oil: 437
palm oil: 420
paraffin: 390
peanut oil: 540
phenol: 175
pine oil: 172
propane: −100
rubber cement: 50
soybean oil: 540

flashpoint of various materials — *continued*

stearic acid: 385
Stoddard solvent: 105
styrene: 90
sulfur: 405
toluene: 40

tung oil: 552
turpentine: 95
whale oil: 446
xylene: 77
xylol: 63

FLUID FLOW: where

a = angle of triangular weir, degrees
A = area, pipe or orifice, in.^{2}
b = width of weir or notch, ft
C_B = coefficient of pipe bend
C_C = coefficient of contraction (reduction)
C_D = coefficient of discharge
C_F = coefficient of pipe fitting
C_V = coefficient of velocity
d = density, slugs/ft^3
D = inside diameter of pipe or orifice, in.
D_M = diameter of venturi meter neck, in.
eff = efficiency, decimal
f = friction factor
g = gravitational acceleration = 32.174 ft/sec^2
h = head, ft (fluid)
h_E = head, energy, ft (fluid)
h_F = head loss, friction, ft (fluid)
h_L = head loss, ft (fluid)
h_M = head, meter static pressure, ft (fluid)
h_P = head, potential energy, ft (fluid)
h_S = head, static pressure, ft (fluid)
h_T = head, total pressure, ft (fluid)
h_V = head, velocity pressure, ft (fluid)
hp = horsepower
L = length of pipe, ft
L_E = length equivalent of pipe fittings, ft
N_R = Reynold's number
p = pressure, psig
$p_1 - p_2$ = pressure drop from point 1 to point 2, psig
Q = actual fluid flow, gal/min

FLUID FLOW — *continued*

Q_T = theoretical flow, gal/min
Q_w = flow rate by weight, lb/hr
s = stress (fiber), psi
S = specific gravity of fluid
t = pipe thickness, in.
v = specific volume, ft³/lb
V = velocity of fluid, ft/sec
V_L = velocity of fluid, large pipe, ft/sec
V_S = velocity of fluid, small pipe, ft/sec
w = specific weight of fluid, lb/ft³
μ = absolute viscosity of fluid, cP
ν = kinematic viscosity of fluid, cSt

absolute viscosity: VISCOSITY, ABSOLUTE
absolute viscosity of various liquids:

viscosity (absolute) of various liquids   *

area, cross-section of pipe or orifice:

$$A = 0.7854D^2$$

Bernoulli's equation: conservation of energy equation, states that the total energy is constant in the steady flow of a frictionless and noncompressible fluid, whereby

potential energy + pressure energy + velocity energy equals a constant,

$$h_P + h_S + h_V - \text{constant}$$

or between two points of a pipe (point 1 and point 2), disregarding friction losses, the total energy at point 1 equals the total energy at point 2,

$$h_{P1} + h_{S1} + h_{V1} = h_{P2} + h_{S2} + h_{V2}$$

or between two points of a pipe (point 1 and point 2), including energy added and friction and other losses between points 1 and 2,

$$h_{P1} + h_{S1} + h_{V1} + h_E = h_{P2} + h_{S2} + h_{V2} + h_F + h_L$$

FLUID FLOW — *continued*

Blasius formula: friction factor   *
coefficient of contraction (reduction):

orifice,

$$C_C = \text{vena contracta area/orifice area}$$

pipe diameters, where

$$R_A = \text{small pipe area/large pipe area}$$

R_A	0.1	0.2	0.3	0.4	0.5
C_C	0.363	0.339	0.308	0.268	0.219

R_A	0.6	0.7	0.8	0.9	1.00
C_C	0.164	0.105	0.053	0.015	0.000

coefficient of cubical expanation: HEAT TRANSFER
coefficient of discharge (C_D):

nozzle,

$$C_D = 0.9 \text{ for rounded outlet, approx.}$$
$$= 0.8 \text{ for square edge outlet, approx.}$$

orifice,

$$C_D = C_C C_V$$
$$= 0.95 \text{ for rounded edge, approx.}$$
$$= 0.60 \text{ for sharp or square edge, approx.}$$

pipe outlet,

$$C_D = 0.9 \text{ for rounded outlet, approx.}$$
$$= 0.8 \text{ for square edge outlet, approx.}$$

venturi meter, as designed by manufacturer or found in venturi meter charts.
weir,

$$\text{rectangular notch} = 0.60 \text{ approx.}$$
$$\text{triangular notch} = 0.55 \text{ approx.}$$

FLUID FLOW — *continued*

coefficient of pipe bends (C_B):

for conversion to head loss (h_L), where

$$R_B = \text{pipe diameter/bend radius}$$

R_B	0.2	0.4	0.6	0.8	1.0
C_B	0.131	0.138	0.158	0.206	0.294

coefficient of pipe fittings (C_F):

for conversion to pipe length equivalent (L_E),

elbow, 90°: 0.8	valves, open, angle: 4.0
45°: 0.55	check: 2.3
sweep: 0.63	gate: 0.17
tee: 1.63	globe: 8.3

coefficient of velocity (C_V) orifice:

may be found in manufacturer's manual,

$$C_V = \text{actual velocity/theoretical velocity}$$

coefficient of viscosity: VISCOSITY, ABSOLUTE
coefficient of viscosity of various liquids: viscosity (absolute) of various liquids *
conservation of energy: Bernoulli's equation *
density: mass per unit volume,

English system, slugs/ft^3,

$$d = w/y - 0.0311w$$

Metric system, grams/cm^3

diameter of pipe required: pipe *
dynamic viscosity: VISCOSITY, ABSOLUTE
efficiency:

$$\text{eff} = \text{output/input}$$

FLUID FLOW — *continued*

flow types: pipe   *

friction factor (f): may be obtained from Moody diagram or approximated for a smooth clean pipe as

$$f = 0.02 \text{ approx.}$$

laminar flow, $N_R < 2000$,

$$f = 64/N_R \text{ (Hagen–Poiseuille law)}$$

transitional flow, $2000 < N_R < 4000$,

$$f = 0.3164/(N_R)^{1/2} \text{ (Blasius formula)}$$

turbulent flow, $N_R > 4000$,

$$f = 0.3164/(N_R)^{1/4} \text{ (Blasius formula)}$$

friction loss: head loss, friction   *

Hagen–Poiseuille law: friction factor, laminar flow   *

head, converted to pressure:

$$p = 0.0069hw$$
$$= 0.433hS$$

head, energy added (as pump):

$$h_E = 2.47 \text{ E} + 05 \text{ (hp)(eff)}/wQ$$

head, meter static pressure (h_M): static pressure head in neck of a meter, as per manufacturer's specification.

head, potential energy (h_P): head pressure at a point of discharge of a fluid, produced by the height of the fluid above that point, and is equal in all directions at that point.

head, static pressure (h_S): hydrostatic head, pressure energy head on the system based on pressure and the weight of the fluid,

$$h_S = 144p/w$$
$$= 2.307p \text{ (for water)}$$
$$= 2.307p/S \text{ (for other fluids)}$$

FLUID FLOW — *continued*

head, total pressure:

$$h_T = h_V + h_S + h_L + h_F$$

head, velocity pressure: kinetic energy, energy of a fluid due to its velocity,

$$h_V = h_T - h_S = V^2/2g$$
$$= 0.0155V^2 = 0.00259Q^2/D^4$$

head loss, converted to pressure drop:

$$p_1 - p_2 = 0.0069wh_L$$

head loss, friction (based on pipe length and fittings):

$$h_F = 0.1865f(L + L_E)V^2/D$$
$$= 0.0311f(L + L_E)Q^2/D^5$$

head loss, other than pipe friction:

pipe bends,

$$h_L = 0.0155C_B V^2$$

pipe diameter increase,

$$h_L = 0.0155(V_L^2 - V_S^2)$$

pipe diameter reduction (contraction),

$$h_L = 0.0155C_C V_L^2$$

pipe entrance,

$$h_L = 0.0078V^2$$

horsepower:

$$\text{hp} = 5.83\,\text{E} - 04\;Q(p_1 - p_2)$$

hydrostatic head: head, static pressure *
kinematic viscosity: VISCOSITY, KINEMATIC
kinematic viscosity of various liquids: viscosity (kinematic) of various liquids *

FLUID FLOW — *continued*

kinetic energy: head, velocity pressure   *
laminar flow: LAMINAR FLOW
length equivalent of pipe fittings:

$$L_E = 2.75 C_F D$$

lohm: LOHM
Moody diagram: MOODY DIAGRAM
nozzle flow:

by volume,

$$Q = 3.12 V A = 2.448 D^2 V$$
$$= 19.63 D^2 C_D h_S^{1/2} = 29.82 D^2 C_D p^{1/2}$$

by weight,

$$Q_W = 25 A V w = 19.63 D^2 V w$$

nozzle velocity of flow:

$$V = 0.3208 Q/A = 0.4085 Q/D^2$$
$$= 8.02 C_D h^{1/2} = 12.18 C_D p^{1/2}$$

orifice: specific size of an opening in a vessel for the purpose of metering the flow of a fluid from the vessel, based on the potential head maintained above the orifice.

flow, actual,

$$Q = 3.12 A V = 2.448 D^2 V = 25 A C_D h_P^{1/2}$$
$$Q w = 25 A V w = 19.63 D^2 V w$$

flow, theoretical,

$$Q_T = 25 A h_P^{1/2}$$

velocity, actual,

$$V = 8.02 C_V h_P^{1/2} = 0.3208 Q/A = 0.4085 Q/D^2$$

velocity, theoretical,

$$V = (2 g h_P)^{1/2} = 8.02 h_P^{1/2}$$

FLUID FLOW — *continued*

pipe:

 diameter required,

$$D = 0.5(Q^2 fL/h_F)^{1/5}$$

flow, laminar (Hagen–Poiseuille),

$$Q = 3664D^4(p_1 - p_2)/\mu L$$
$$= 19.63D^2 C_D h_S^{1/2} = 29.82D^2 C_D p^{1/2}$$

 flow, by volume,

$$Q = 3.12AV = 2.448D^2 V = 5.67(D^5 h_F/fL)^{1/2}$$

 flow, by weight,

$$Q_W = 25AVw = 19.63D^2 Vw - 8.02wQ$$

flow types, based on Reynold's number (N_R):

laminar flow, $N_R < 2000$
transitional flow, $2000 < N_R < 4000$
turbulent flow, $N_R > 4000$

stress on pipe wall,

$$s = 0.5pD/t$$

 thickness of pipe wall required,

$$t = 0.5pD/s$$

 velocity, of flow,

$$V = 8h_S^{1/2} = 12.2p^{1/2}$$
$$= 0.408Q/D^2 = 0.3208Q/A$$
$$= 1497D^2(p_1 - p_2)/\mu L$$

Poiseuille's law: in the laminar flow of a fluid flowing in a pipe, the pressure loss varies directly with the length of the pipe, fluid velocity, fluid viscosity, and inversely as the fourth power of the pipe diameter.
potential energy head: head, potential energy   *

FLUID FLOW — *continued*

pressure converted to head:

$$h = 144p/w = 2.307p/S$$

pressure drop converted to head loss:

$$h_L = 144(p_1 - p_2)/w = 2.307(p_1 - p_2)/S$$

pressure drop due to friction, based on pipe length:

$$p_1 - p_2 = 0.001295V^2wfL/D$$
$$= 0.0808V^2fLS/D$$

pressure energy head: head, static pressure ✳
pump formulas: PUMP FORMULAS FOR FLUIDS
Reynold's number (N_R): dimensionless number used in fluid flow piping calculations,

$$N_R = 124wVD/\mu = 50.65wQ/D\mu = 3162QS/D\mu$$
$$N_R = 7742VD/v = 3162Q/Dv$$

specific viscosity: viscosity, specific ✳
specific volume: reciprocal of specific weight,

$$v = 1/w$$

specific weight: reciprocal of specific volume,

$$w = 1/v$$

specific weight of various liquids: WEIGHT (SPECIFIC) OF VARIOUS LIQUIDS
static head: head, static pressure ✳
static pressure: pressure of a fluid, in a pipe, that tends to burst or collapse the pipe,

$$p = 0.0069wh_S = 0.433h_S S$$

static pressure head: head, static pressure ✳
tank discharge, open pipe in wall of tank:

flow,

$$Q = 3.12AV = 2.448D^2V$$

FLUID FLOW — *continued*

velocity,

$$V = 8.02h_P^{1/2}$$

total pressure head: head, total pressure　*
transitional flow: TRANSITIONAL FLOW
turbulent flow: TURBULENT FLOW
velocity energy: head, velocity pressure　*
velocity pressure: in the flow of the fluid, the pressure attributed to the velocity of the fluid.
velocity pressure head: head, velocity pressure　*
vena contracta: the smallest cross-section in a stream of fluid as it is discharged through an orifice of the tank containing the fluid.
venturi meter: an instrument installed in a pipeline for the purpose of measuring the rate of flow of a fluid through the pipe. The venturi is a gradual reduction of the size of the pipe to a "neck" and then returned to the original size of the pipe. It is especially designed to eliminate any turbulence through the metering section. As velocity increases through the neck of the meter, static pressure is reduced at that point and the resulting pressure drop is used to determine the flow of the fluid.

venturi meter flow measurement,

$$Q = \frac{19.64 C_D (D_M D)^2 (h_S - h_M)^{1/2}}{(D^4 - D_M^4)^{1/2}}$$

viscosity: VISCOSITY
VISCOSITY (ABSOLUTE) OF VARIOUS LIQUIDS (cP, 70°F)

acetic acid: 1.155	carbon tetrachloride: 0.86
acetone: 0.316	castor oil: 600
aniline: 3.35	crude oil: 6.5
benzene: 0.60	ether: 0.25
benzine: 0.63	ethyl alcohol: 1.15
calcium chloride (20%): 2.3	ethylene glycol: 23

FLUID FLOW — *continued*

fuel oil no. 1: 1.32	octane: 0.51
no. 2: 2.32	oil, heavy: 660
no. 4: 13.9	light: 113
no. 5: 48	SAE 10: 130
no. 6: 355	SAE 60: 5400
gasoline: 0.5	phenol: 8.0
glycerine: 700	propane: 0.2
kerosene: 1.8	sodium chloride (20%): 2.0
linseed oil: 44	toluene: 0.55
mercury: 1.6	turpentine: 3
methyl alcohol: 0.56	water: 1.0

viscosity (kinematic) of various liquids (cSt, 70°F)

acetic acid: 1.1	linseed oil: 36
acetone: 0.40	mercury: 0.117
benzene: 0.70	methyl alcohol: 0.71
benzine: 0.854	octane: 0.728
carbon tetrachloride: 0.60	oil, SAE 10: 41
castor oil: 625	SAE 30: 114
crude oil: 8.4	SAE 50: 270
ether: 0.312	SAE 90: 287
ethyl alcohol: 1.49	phenol: 7.47
ethylene glycol: 14.7	propane: 0.222
gasoline: 0.435	toluene: 0.635
glycerine: 550	turpentine: 1.6
kerosene: 2.25	water: 1.00

viscosity, specific: ratio of the absolute viscosity of the fluid to the absolute viscosity of water. In the metric system, specific viscosity equals absolute viscosity.

weir: a notch in the end of a fluid flow box that is set in an open channel to measure the flow rate of the fluid. The rate of flow is determined by the flow rate of the fluid through the box and its head pressure.

weir flow, rectangular notch,

actual,

$$Q = 2400bC_D h_P^{3/2}$$

FLUID FLOW — *continued*

theoretical,

$$Q_T = 2400bh_P^{3/2}$$

weir flow, triangular notch ($a = 90°$),

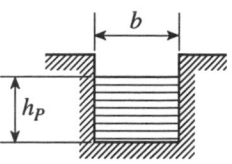

actual,

$$Q = 1920C_D\, h_P^{5/2}$$

theoretical,

$$Q_T = 1920h_P^{5/2}$$

weir flow, triangular notch,

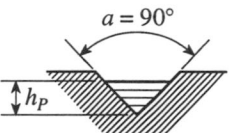

actual,

$$Q = 1920C_D(\tan 0.5a)h_P^{5/2}$$

theoretical,

$$Q_T = 1920(\tan 0.5a)h_P^{5/2}$$

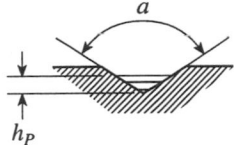

---

flux density: ELECTROMAGNETISM

flywheel effect: energy is stored in the flywheel as its rotating velocity increases. Stored energy is then released with the decrease of velocity of the rotating flywheel as work is performed, to equalize energy and reduce changes in flywheel speed.

FLYWHEEL FORMULAS: where

E = energy required per stroke, ft-lb
hp = horsepower
k = radius of gyration, ft
KE = kinetic energy, ft-lb
N = flywheel speed, rev/min
N_1 = initial speed, rev/min
N_2 = final speed, rev/min
r = radius of flywheel, ft
S = number of strokes per minute

FLYWHEEL FORMULAS — *continued*

t = time, sec
T = torque applied, ft-lb
W = weight of flywheel, lb
w_1 = initial angular velocity, rad/sec
w_2 = final angular velocity, rad/sec

horsepower: hp = $ES/33,000$
kinetic energy, rim type:

$$KE = 1.704 \text{ E} - 04 \; Wk^2N^2$$

kinetic energy, solid wheel:

$$KE = 8.52 \text{ E} - 05 \; Wr^2N^2$$

time required to change speed:

$$t = 0.003256Wk^2(N_2 - N_1)/T$$
$$= 0.0311Wk^2(w_2 - w_1)$$

---

foot =

0.00139 cable length	5.486 E − 05 league (naut)
0.01515 chain (Gunter)	6.313 E − 05 league (stat)
0.01 chain (Ramden)	1 link (engineer)
30.48 cm	1.515 links (Gunter)
0.667 cubit	1 link (Ramden)
0.03048 dam	0.3048 m
3.048 dm	304,800 μm
0.2667 ell	1.646 E − 04 mi (naut)
0.1667 fathom	1.894 E − 04 mi (stat)
0.001515 furlong	12,000 mils
0.003048 hm	304.8 mm
12 in.	0.0606 rod
3.048 E − 04 km	0.00278 skein
1.646 E − 04 kn	0.3333 yd

foot/hour =

30.48 cm/h	3.048 E − 04 km/h
0.508 cm/min	5.08 E − 06 km/min
0.008467 cm/s	1.646 E − 04 kn
0.01667 ft/min	0.3048 m/h
2.778 E − 04 ft/s	8.4667 E − 05 m/s
12 in./h	1.894 E − 04 mi/h
0.20 in./min	3.157 E − 06 mi/min
0.00333 in./s	

foot/minute =

0.508 cm/s	3.048 E − 04 km/min
60 ft/h	0.009875 kn
0.01667 ft/s	18.288 m/h
720 in./h	0.3048 m/min
12 in./min	0.00508 m/s
0.20 in./s	0.01136 mi/h
0.01829 km/h	1.8939 E − 04 mi/min

foot/second =

30.48 cm/s	0.592 kn
60 ft/min	18.288 m/min
720 in./min	0.3048 m/s
12 in./s	0.6818 mi/h
1.0973 km/h	0.01136 mi/min
0.01829 km/min	

foot/second/second =

30.48 cm/s²	0.3048 m/s²
0.03108 gravitational acceleration	2454 mi/h²
constant	0.6818 mi/h/s
1.097 km/h/s	

foot-candle: ILLUMINATION

foot-candle =

1 lm/ft²	10.764 lx
10.764 lm/m²	1.0764 milliphots (mph)

foot-lambert: ILLUMINATION
foot-lambert =

3.426 E − 04 cd/cm²	0.0010764 L
0.3183 cd/ft²	1 lm/ft²
0.00221 cd/in.²	10.764 lm/m²
3.426 cd/m²	1.0764 mL

foot (mercury, 0°C, 32°F) =

0.401 atm	12 in. (mercury)
0.40637 bar	5.8939 lb/in.²

foot (water, 4°C, 39.2°F) =

0.0295 atm	2.98598 kPa
0.029859 bar	62.422 lb/ft²
2.242 cm (mercury, 0°C, 32°F)	0.4335 lb/in.²
2.989 E + 04 dyn/cm²	0.0224 m (mercury, 0°C, 32°F)
30.479 g/cm²	22.420 mm (mercury, 0°C, 32°F)
0.8827 in. (mercury, 0°C, 32°F)	0.002989 N/m²
12 in. (water)	2989 Pa
0.03048 kg/cm²	22.40 torr
304.8 kg/m²	

foot (water, 20°C, 68°F) =

0.02944 atm	12 in. (water)
0.02905 bar	2.984 kPa
2.9837 E + 04 baryes	62.317 lb/ft²
30.48 cm (water)	0.4326 lb/in.²
2.9837 E + 04 dyn/cm²	2984 Pa
0.8809 in. (mercury, 0°C, 32°F)	22.38 torr

foot-pound, pound-foot: unit of mechanical energy

foot-pound =

0.001285 Btu	12 in.-lb
1.3558 E + 07 dyn-cm	192 in.-oz
1.3558 E + 07 erg	1.356 J
4.7254 E − 04 ft³-atm	3.24 E − 04 kg-cal
32.174 ft-pdl	0.1383 kg-m
0.324 g-cal	3.766 E − 07 kWh
1.3825 E + 04 g-cm	0.01338 l-atm
5.051 E − 07 hp-h	1.356 N-m
5.12 E − 07 hp-h (metric)	3.766 E − 04 Wh
0.001818 hp-s	1.356 W-s

foot-pound/hour =

0.001285 Btu/h	0.0054 g-cal/min
2.142 E − 05 Btu/min	5.051 E − 07 hp
2.2597 E + 05 erg/min	5.12055 E − 07 hp (metric)
0.01667 ft-lb/min	3.766 E − 07 kW
2.7778 E − 04 ft-lb/s	3.766 E − 04 W

foot-pound/minute =

0.077156 Btu/h	3.03 E − 05 hp
0.001285 Btu/min	2.3036 E − 06 hp (boiler)
2.143 E − 05 Btu/s	3.0723 E − 05 hp (metric)
2.2597 E + 05 erg/s	0.022597 J/s
60 ft-lb/h	3.24 E − 04 kg-cal/min
0.01667 ft-lb/s	2.26 E − 05 kW
0.0054 g-cal/s	0.0226 W

foot-pound/pound =

0.001285 Btu/lb	0.002989 J/g
7.144 E − 04 g-cal/g	3.048 E − 04 kg-m/g
5.050 E − 07 hp-h/lb	8.303 E − 10 kWh/g

foot-pound/second =

4.625 Btu/h	0.001843 hp (metric)
0.0771 Btu/min	1.35582 J/s
0.001285 Btu/s	0.01943 kg-cal/min
1.3558 E+07 erg/s	3.24 E−04 kg-cal/s
60 ft-lb/min	0.1383 kg-m/s
0.32405 g-cal/s	0.001356 kW
1.3825 E+04 g-cm/s	1.356 W
0.001818 hp	

foot-poundal =

3.9968 E−05 Btu	429.712 g-cm
0.41589 cm³-atm	1.57 E−08 hp-h
4.214 E+05 dyn-cm	0.04214 J
4.214 E+05 erg	1.007 E−05 kg-cal
1.4687 E−05 ft³-atm	0.004297 kg-m
0.031081 ft-lb	1.171 E−08 kWh
0.01007 g-cal	4.1588 E−04 l-atm

foot-poundal/minute =

7.02 E−07 kW
7.02 E−04 W

foot square/second: SQUARE FOOT/SECOND
force: energy of one body that causes another body to move or to change the movement or shape of the body acted on, or cause that body to accelerate; equals mass times acceleration. Force units are

English gravitational system, lb
Metric absolute system, dyn

force formulas: LINEAR MOTION, STATICS
formulas, chemical: CHEMICAL FORMULAS FOR VARIOUS COMPOUNDS
Fr: francium (element)
freezing point: temperature and pressure at which the liquid and solid phases of a material are at equilibrium.

freezing point of various materials (°F):

acetic acid: 62	hydrogen iodide: −60
alcohol: −170	hydrogen sulfide: −122
ammonia: −108	iodine: 235
argon: −308	isobutane: −229
benzene: 42	linseed oil: −4
butane: −217	mercury: −38
carbon dioxide: −70	methane: −297
carbon disulfide: −169	methyl acetylene: −153
carbon monoxide: −341	methyl alcohol: −144
carbon tetrachloride: −9	methyl chloride: −144
castor oil: 14.1	naphthalene: 176
chlorine: −150	nitric acid: −42
ether: −180	nitric oxide: −262
ethyl alcohol: −175	nitrous oxide: −131
ethyl chloride: −217.7	octane: −70.2
ethylene: −273	oxygen: −361
ethylene glycol: 8.6	phenol: 104
fluorine: −369	phosgene: −198
freon F12: −252	propane: −305
freon F22: −256	sea water: 27.5
glycerine: 17	sulfur dioxide: −104
helium: −458	toluene: −139
hydrogen: −435	turpentine: −75
hydrogen bromide: −124.6	vinyl chloride: −255
hydrogen chloride: −173	water: 32
hydrogen fluoride: −117	

frequency: ELECTRICAL

frequency of various waves: WAVE FREQUENCIES OF VARIOUS WAVES

FRICTION: a mechanical term for the resisting force to motion (or sliding) between the contact surfaces of two bodies in motion or between one body in motion and its contact surface.

angle of repose: the angle of inclination from a horizontal surface at which the material slides by its own weight.

FRICTION — *continued*

angle of repose for various materials (degrees):

clay, compact: 20–25   rip rap: 45
    soft: 10   salt: 36
coal: 25–35   sand, dry: 25–35
coke: 30–45   moist: 30–45
gravel, round: 30   wet: 20–40
    sharp: 40   silt, compact: 25–40
iron ore: 35–45   loose: 20–40
masonry rubble: 35

coefficient of friction: friction coefficient *

friction coefficient; coefficient of friction: a measure of the resisting force to motion (or sliding) between the contact surfaces of two bodies or between one body in motion and its contact surface.

friction force: the force required to overcome friction between two bodies in contact to set one body in motion.

Rolling friction: the ratio of torque applied at the center of a wheel to the load on the wheel, as required to roll the wheel at a steady speed, where

c = rolling friction coefficient, in.
F = force at center of wheel and parallel to the line of travel, lb
R = radius of wheel, in.
T = torque, in.-lb
W = load on wheel, lb

force required,

$$F = cW/R$$

rolling friction coefficient,

$$c = T/W = FR/W$$

Rolling friction coefficients (c) **(in.):**

iron roller/asphalt: 0.144   steel roller/steel: 0.02
iron roller/iron: 0.020   wood roller/wood: 0.005
iron roller/wood: 0.22

FRICTION — *continued*

Sliding friction: the pulling force required to continue to slide a body in motion on a horizontal surface. To start the body in motion requires a greater pulling force and is considered the static friction force, where

f = sliding friction coefficient
F = friction force, lb
N = normal force, lb
P = pulling force, lb
W = weight of body, lb

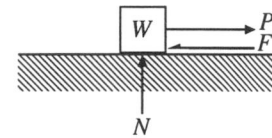

friction force,

$$F = P = fN$$

normal force, $N = W$
sliding friction coefficient,

$$f - F/N = P/N = P/W$$

pulling force, $P = F$

Sliding friction coefficients (f):

brass/dry wood: 0.48	iron/dry wood: 0.49
bronze/bronze: 0.20	iron/greased wood: 0.20
bronze/iron: 0.21	iron/wet wood: 0.25
bronze/greased iron: 0.09	oak/oak, against grain: 0.19
iron/iron: 0.40	oak/oak, with grain: 0.48
iron/greased iron: 0.09	steel/steel: 0.09
Iron/wet Iron: 0.31	

Static friction: the ratio of static friction force (F) to normal pressure (N) between two bodies in motion, measured as the tangent of the minimum angle of a surface that is required to move the body on that surface, where

a = angle of surface with horizontal, degrees
f = static friction coefficient
F = friction force, lb
N = normal pressure, lb
P = pushing force, lb
W = weight of body, lb

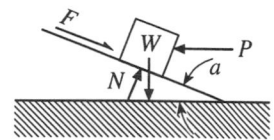

FRICTION — *continued*

friction force,

$$F = fN$$

normal force,

$$N = W/(\cos a - f \sin a)$$

pushing force,

$$P = N(\sin a + f \cos a)$$

static friction coefficient,

$$f = \tan a = F/N$$

Static friction coefficients (f):

iron/greased iron: 0.16	oak/oak with grain: 0.62
iron/wet oak: 0.65	steel/steel: 0.15
oak/oak against grain: 0.43	

---

friction factor: FLUID FLOW
friction loss: FLUID FLOW
fuel/air ratio for combustion: AIR/FUEL RATIO FOR COMBUSTION
fuel heating values: HEAT VALUE OF VARIOUS GASES;
                             HEAT VALUE OF VARIOUS LIQUIDS;
                             HEAT VALUE OF VARIOUS MATERIALS.

furlong =

2.01168 E+04 cm	201.168 m
10 chains (Gunter)	0.1086 mi (nautical)
6.6 chains (Ramden)	0.125 mi (statute)
660 ft	40 rods
7920 in.	220 yd

fusion heat: HEAT TRANSFER

G

Ga: gallium (element)
gallon (dry) =

0.0381 bbl (dry)
0.0369 bbl (liquid)
0.125 bushel (level)
4405 cm³
4.405 dm³
0.1556 ft³
0.96932 gal (Imperial)
1.164 gal (liquid)
268.803 in.³
0.004405 kl

4.405 l
0.004405 m³
148.95 oz (liquid)
0.5 peck
8 pt (dry)
4 qt (dry)
4.564 qt (liquid)
0.004405 stere
0.00576 yd³

gallon (Imperial) =

0.12896 bushel
4546 cm³
0.16054 ft³
1.0317 gal (dry)
1.20095 gal (liquid)
277.27 in.³
0.004546 kl
4.546 l

0.004546 m³
4546 ml
153.72 oz (liquid)
4.1267 qt (dry)
4.80128 qt (liquid)
0.004546 stere
0.005946 yd³

gallon (liquid) =

3.069 E − 06 acre-ft	231 in.^{3}
3.683 E − 05 acre-in.	0.003785 kl
0.03175 bbl (liquid)	3.785 l
0.02381 bbl (petroleum)	0.003785 m^3
0.1074 bushel (level)	3785 ml
3785 cm^3	61,440 minims
0.3785 dal	128 oz (liquid)
3.7854 dm^3	0.43 peck
0.13368 ft^3	6.875 pt (dry)
0.859 gal (dry)	8 pt (liquid)
0.83267 gal (Imperial)	3.4378 qt (dry)
32 gills	4 qt (liquid)
0.03785 hl	0.003785 stere
0.01587 hogshead	0.004951 yd^3

gallon/hour =

3.069 E − 06 acre-ft/h	3.7854 l/h
0.13368 ft^3/h	6.31 E − 05 m^3/min
0.002228 ft^3/min	8.252 E − 05 yd^3/min
3.713 E − 05 ft^3/s	

gallon/minute =

0.002212 acre-in./h	3.785 l/min
63.09 cm^3/s	0.06309 l/s
8.021 ft^3/h	0.2271 m^3/h
0.13368 ft^3/min	0.003785 m^3/min
0.002228 ft^3/s	6.309 E − 05 m^3/s
60 gal/h	63.09 ml/s
0.01667 gal/min	

gallon/second =

3785 cm^3/s	60 gal/min
481.25 ft^3/h	227.12 l/min
8.02 ft^3/min	3.7854 l/s
0.13368 ft^3/s	3785 ml/s
3600 gal/h	0.2971 yd^3/min

galvanic cell: ELECTROLYSIS

galvanic corrosion: corrosion caused by the electrical current produced from the chemical reaction between two joined dissimilar metals in the presence of a liquid (electrolyte).

galvanic series: ELECTROLYSIS

gamma: equals $1 E-09$ g

gas, gram molecular weight: weight of 1 mole of gas occupying 22.4 l.

gas, ideal: perfect gas; gas that behaves according to the Gas Laws.

gas, natural: NATURAL GAS

gas, standard conditions: gas at 1 atm and 32°F.

gas compression formulas: AIR/GAS COMPRESSION FORMULAS

gas constant (R): for any perfect gas, where

M = molecular weight of the gas
P = absolute pressure, lb/ft^2
R = gas constant, ft-lb/lb/°R
"R" = universal gas constant, 1545.3 ft-lb/lb-mol-°R
T = absolute temperature, °R
v = specific volume, ft^3/lb

$$\text{gas constant, } R = \text{"R"}/M = 1545.3/M = Pv/T$$

gas constant, universal ("R"):

1.986 Btu/lb-mol-°R	1545.3 ft-lb/lb-mol-°R
82.07 cm^3-atm/g-mol-K	1.983 g-cal/mol/K
8.31 E+07 erg/g-mol-K	8.31 J/g-mol-K
21.85 ft^3-in. (mercury)/lb-mol-°R	0.08207 l atm/g-mol-K

gas constants (R) for various gases (ft-lb/lb-mol-°R)

acetylene: 59.4	chlorine: 21.79
air: 53.35	ethane: 51.43
ammonia: 90.73	ethylene: 55.13
argon: 38.65	freon F12: 12.78
butane: 26.61	helium: 386.3
carbon dioxide: 35.10	hydrogen: 767.0
carbon disulfide: 20.3	hydrogen sulfide: 45.35
carbon monoxide: 55.19	isobutane: 25.79

gas constants (R) for various gases — *continued*

methane: 96.40	propane: 35.07
neon: 76.56	steam: 85.58
nitrogen: 55.15	sulfur: 24.00
nitrous oxide: 35.11	sulfur dioxide: 24.12
octane: 13.54	water vapor: 85.58
oxygen: 48.29	

GAS FLOW FORMULAS: where

A = area of pipe, in.^{2}
acfm = actual ft^3/min
C_F = coefficient of pipe fitting
d = density, slugs/ft^3 or grams/cm^3
D = inside diameter of pipe, in.
f = friction factor
g = gravitational acceleration = 32.174 ft/sec^2
h = head, in. (water)
h_F = head loss, friction, in. (water)
h_L = head loss, in. (water)
h_P = head, potential energy, in. (water)
h_S = head, static pressure, in. (water)
h_T = head, total pressure, in. (water)
h_V = head, velocity pressure, in. (water)
L = length of pipe, ft
L_E = length equivalent of pipe fittings, ft
N_R = Reynold's number
p = gauge pressure, psig
p_1 = initial pressure, psig
p_2 = final pressure, psig
P = absolute pressure, psia
P_1 = initial absolute pressure, psia
P_2 = final absolute pressure, psia
Q = actual flow rate, acfm
Q_W = flow rate by weight, lb/min
R = gas constant, ft-lb/lb/$^\circ$R
scfm = standard ft^3/min
t = temperature, $^\circ$F
T = absolute temperature, $^\circ$R
v = volume, ft^3
v_1 = initial volume, ft^3

GAS FLOW FORMULAS — *continued*

v_2 = final volume, ft^3
V = velocity of gas, ft/min
w = specific weight of gas, lb/ft^3
W = weight of gas, lb
μ = absolute viscosity, cP
v = kinematic viscosity, cSt

(⊗) absolute pressure, psia:

$$P = p + 14.696$$
$$= 0.433 \,[\text{ft (water) gauge} + 33.898]$$
$$= 0.491 \,[\text{in. (mercury) gauge} + 29.92]$$
$$= 0.036 \,[\text{in. (water) gauge} + 406.77]$$

(⊗) absolute temperature, °R:

$$T = t + 459.69$$

absolute viscosity: VISCOSITY, ABSOLUTE
absolute viscosity of various gases: viscosity (absolute) of various gases   *
actual cubic feet per minute (acfm): actual flow rate by volume.
area cross-section of pipe:

$$A = 0.7854D^2$$

Bernoulli's equation: conservation of energy; total energy is constant in the steady flow of an incompressible gas whereby potential energy + pressure energy + velocity energy equals a constant.

$$h_P + h_S + h_V = \text{constant}$$

or between points 1 and 2,

$$h_{P1} + h_{S1} + h_{V1} = h_{P2} + h_{S2} + h_{V2}$$

coefficient of cubical expansion: HEAT TRANSFER
coefficient of pipe fittings (C_F): for conversion to pipe length equivalent (L_E),

elbow, 90°: 0.8   valves (open), angle: 4.0
     45°: 0.55   check: 2.3
     sweep: 0.63   gate: 0.17
tee: 1.63   globe: 8.3

GAS FLOW FORMULAS — *continued*

coefficient of viscosity: VISCOSITY, ABSOLUTE
coefficient of viscosity of various gases: viscosity (absolute) of various gases *

conservation of energy: Bernoulli's equation   *

density: mass per unit volume,

English system, slugs/ft^3

$$d = w/g = 0.0311w = 4.476P/RT$$

Metric system, grams/cm^3

dynamic viscosity: VISCOSITY, ABSOLUTE

(Ⓢ) **flow rate, laminar flow:**

by volume,

$$Q = 0.0069AV = 0.00545D^2V$$

by weight,

$$\begin{aligned} Q_W &= wQA \\ &= 0.0069wVA \\ &= 0.00545D^2Vw \end{aligned}$$

(Ⓢ) **flow rate, turbulent flow:**

by volume,

$$Q = 19.3[R(P_1^2 - P_2^2)D^5/fTL]^{1/2}$$

by weight,

$$Q_W = 77[(P_1^2 - P_2^2)D^5/fRTL]^{1/2}$$

flow types, based on Reynold's number:

laminar flow, $N_R < 2000$
transitional flow, $2000 < N_R < 4000$
turbulent flow, $N_R > 4000$

friction factor (f): may be obtained from a Moody diagram or other charts.

friction loss: head loss, friction   *

GAS FLOW FORMULAS — *continued*

gas constants (R): GAS CONSTANTS (R)

head, converted to pressure in psi: $p = 0.036h$

head, static pressure: static pressure energy head on the system based on pressure and weight of the gas, as measured with a manometer perpendicular to the flow.

$$h_S = h_T - h_V$$

head, total pressure:

$$h_T = h_S + h_V + h_L + h_F$$

head, velocity pressure: kinetic energy, energy of the gas due to its velocity,

$$h_V = h_T - h_S$$
$$= 8.3\,\text{E}-07\ V^2 w$$

head loss, friction (based on pipe length and fittings):

$$h_F = 9.957\,\text{E}-06\,f(L + L_E)V^2 w/D$$
$$= 12 f(L + L_E)h_V/D$$

kinematic viscosity: VISCOSITY, KINEMATIC

kinematic viscosity of various gases: viscosity (kinematic) of various gases *

kinetic energy: head, velocity pressure *

laminar flow: LAMINAR FLOW

length equivalent (L_E) of pipe fittings:

$$L_E = 2.750 C_F D$$

manometer: MANOMETER

Moody diagram: MOODY DIAGRAM

pipe fitting coefficients: coefficients of pipe fittings *

pitot tube: PITOT TUBE

potential energy (h_P): energy stored due to elevation of the system above datum.

pressure drop, converted to head loss:

$$h_L = 27.76(p_1 - p_2)$$

GAS FLOW FORMULAS— *continued*

 **pressure drop due to friction:**

$$p_1 - p_2 = 3.6 \text{ E}-07 \, fw(L + L_E)V^2/D$$

Reynold's number:

$$N_R = 2.07VDw/\mu$$
$$= 129VD/v$$

specific volume: reciprocal of specific weight,

$$v = 1/w$$

specific weight: reciprocal of specific volume,

$$w = 1/v$$
$$= 70.729P/RT, \text{ if } P \text{ is in inches (mercury)}$$
$$= 5.1868P/RT, \text{ if } P \text{ is in inches (water)}$$
$$= 144P/RT, \text{ if } P \text{ is in psia}$$

specific weight of various gases: WEIGHT (SPECIFIC) OF VARIOUS GASES
standard conditions:

$$\text{pressure} = 1 \text{ atm}$$
$$= 33.898 \text{ ft (water)}$$
$$= 29.92 \text{ in. (mercury)}$$
$$= 14.696 \text{ psia}$$
$$\text{temperature} = 70°F = 529.69°R$$

standard cubic feet per minute (scfm): flow rate at standard conditions, converted from actual flow rate,

$$\text{scfm} = 36(\text{acfm})P/T$$

static head: head, static pressure *
static pressure: pressure that tends to burst or collapse the pipe.
static pressure head: head, static pressure *
total pressure head: head, total pressure *
transitional flow: TRANSITIONAL FLOW
turbulent flow: TURBULENT FLOW

GAS FLOW FORMULAS — *continued*

 velocity of flow:

$$V = 1098(h_V/w)^{1/2}$$
$$= 144Q/A = 183.35Q/D^2$$
$$= 144Q_w/wA = 183.35Q_wD^2w$$

velocity pressure: in the flow of a gas, the pressure attributed to the velocity of the gas.

velocity pressure head: head, velocity pressure   *

viscosity: VISCOSITY

viscosity (absolute) of various gases (cP):

acetylene: 0.010	freon F22: 0.013
air: 0.018	helium: 0.0194
ammonia: 0.01	hydrogen: 0.0089
argon: 0.023	methane: 0.011
butane: 0.007	neon: 0.032
carbon dioxide: 0.0142	nitrogen: 0.018
carbon monoxide: 0.0176	oxygen: 0.020
chlorine: 0.014	propane: 0.08
ethane: 0.095	steam: 0.011 (212°F)
fluorine: 0.024	water vapor: 0.011 (212°F)
freon F12: 0.013	

viscosity (kinematic) of various gases (cSt):

air: 15.79	hydrogen: 111.5
carbon dioxide: 8.2	oxygen: 15.7
helium: 121	steam: 1.77 (212°F)

volume change due to temperature: at constant pressure, the volume of a gas expands directly with temperature change.

$$v_2 = v_1(T_2/T_1)$$

weight, specific: specific weight   *

GAS LAWS: where

P_1, P_2 = absolute pressure, initial and final, psia
T_1, T_2 = absolute temperature, initial and final, °R
V_1, V_2 = volume, initial and final, ft^3

Amagat's law: the sum of the partial volumes of gases, at a specific pressure and temperature, equals the total volume at that same pressure and temperature.

Avogadro's law: equal volumes of gases at the same pressure and temperature contain the same number of molecules.

Boyle's law: at constant temperature, the volume of a fixed weight for a given gas is inversely proportional to the pressure under which it is measured, as pressure multiplied by volume equals a constant,

$$P_1 V_1 = P_2 V_2 = \text{constant}$$

Boyle's and Charles' laws combined: at constant volume, the pressure of a gas varies directly as its absolute temperature,

$$(P_1 V_1)/T_1 = (P_2 V_2)/T_2 = \text{constant}$$

or

$$P_1/T_1 = P_2/T_2 = \text{constant}$$

because $V_1 = V_2$.

Charles' law for a perfect gas: under constant pressure, the volume of a fixed weight of a gas is directly proportional to its absolute temperature,

$$V_1/T_1 = V_2/T_2$$

Dalton's gas law: each gas in a gaseous mixture exerts a partial pressure according to the volume of the mixture. The sum of the partial pressures exerted by the individual component gases equals the total pressure of the gaseous mixture.

gas mole = 6.023 E + 23 molecules

gauge pressure (psig): reading above or below atmospheric pressure; pressure as obtained directly with the use of an ordinary pressure gauge.

gauss: ELECTROMAGNETISM
gauss =

1 line/cm^2	3.336 E − 11 statweber/cm^2
6.452 lines/in.^{2}	1 E − 08 Wb/cm^2
1 Mx/cm^2	6.4516 E − 08 Wb/in.^{2}
6.4516 Mx/in.^{2}	1 E − 04 Wb/m^2

gauss/oersted: ELECTROMAGNETISM
gauss/oersted = 1.257 E − 06 H/m
gauss-square centimeter: MAXWELL
Gd: gadolinium (element)
Ge: germanium (element)
GEAR FORMULAS—SPUR GEARS ($14\frac{1}{2}$ or 20°): where:

A = addendum
B = dedendum
c = clearance at base of tooth
C = center-to-center distance
cp = circular pitch
D = outside diameter
n = number of teeth
N = speed, rev/min
P = diametral pitch
pd = pitch diameter
R = ratio
T = thickness of tooth at pitch circle
W = depth of tooth, total
wd = depth of tooth, working

addendum: length of tooth from pitch diameter to outside diameter,

$$A = 1/P = 0.3183(\text{pd})$$

center-to-center distance: of two gears (1 and 2) in mesh,

$$\begin{aligned} C &= 0.5 \, [(\text{pd})_1 + (\text{pd})_2] \\ &= 0.5(n_1 + n_2)/P \\ &= 0.159(n_1 + n_2)(\text{cp}) \end{aligned}$$

GEAR FORMULAS — *continued*

circular pitch: distance along pitch circle from point on one tooth to corresponding point on adjacent tooth,

$$\text{cp} = 3.1416/P$$
$$= 3.1416(\text{pd})/n$$

clearance: distance between base of tooth and top of engaged tooth,

$$c = 0.157/P = 0.05(\text{cp})$$

dedendum: length of tooth from pitch diameter to bottom of tooth,

$$B = 1.157/P = 0.3683(\text{cp})$$

depth of tooth, total:

$$W = 2.157/P = 0.6866(\text{cp})$$

depth of tooth, working:

$$\text{wd} = 2/P = 0.6366(\text{cp})$$

diametral pitch: number of teeth per inch of pitch diameter,

$$P = n/(\text{pd}) = 3.1416/(\text{cp})$$

outside diameter: overall diameter of the gear,

$$D = \text{pd} + 2A$$
$$= 0.3183(n + 2)(\text{cp})$$
$$= (n + 2)/P$$

pitch circle: tangent or mating point of gears in mesh.
pitch diameter: diameter of pitch circle,

$$\text{pd} = n/P = 0.3183n(\text{cp})$$

ratio: gear ratio; ratio of the number of teeth in the larger gear (L) to the number of teeth in the smaller gear (S),

$$R = n_L/n_S$$

speed ratio between gears 1 and 2:

$$N_1/N_2 = (\text{pd})_2/(\text{pd})_1$$

GEAR FORMULAS—*continued*

teeth, number:

$$n = P(\text{pd}) = 3.1416(\text{pd})/(\text{cp})$$

tooth thickness at pitch circle:

$$T = 1.5708/P = 0.5(\text{cp})$$

---

gee-pound: SLUG

geometric progression: addition of number (n) of terms (a) where each following term (a) is increased by a multiple of a fixed ratio (r) and the last term is ar^{n-1}, as

$$a + ar + a(r)^2 + a(r^3) + \cdots + a(r)^{n-1}$$

and sum of the terms is

$$T = a(r^n - 1)/(r - 1)$$

giga: prefix, equals $1\ \text{E}+09$

gigameter $= 1\ \text{E}+09$ m

gilbert: ELECTROMAGNETISM

gilbert $=$

0.0796 abampere-turn
0.796 A-turn

gilbert/centimeter: OERSTED

gilbert/maxwell: ELECTROMAGNETISM

gill $=$

118 cm³	1920 minims
32 dr (liquid)	118 ml
0.03125 gal (liquid)	4 oz (liquid)
7.21875 in.³	0.25 pt (liquid)
0.1183 1	0.125 qt (liquid)
1.1829 E − 04 m³	

g-pound; gee-pound: SLUG

grade: portion of a circle; also the up (up grade) or down (down grade) rate of a slope.

grade =

0.0025 circle
0.0025 rev

grade, percent: PERCENT GRADE

grain =

0.31566 carat
6.48 cg
0.03657 dr
0.01667 dr (apoth/troy)
63.546 dyn
0.06480 g
1 gr (apoth/troy)
1 gr (avdp)

6.48 E−05 kg
1.429 E−04 lb
1.736 E−04 lb (apoth/troy)
64.8 mg
0.002286 oz
0.002083 oz(apoth/troy)
0.04167 pwt (troy)
0.05 scruple (apoth)

grain/cubic centimeter =

0.0648 g/cm^3
1 gr/ml

grain/cubic foot = 2.288 g/m^3

grain/gallon =

0.01712 g/l
17.12 g/m^3
0.017118 kg/m^3

142.86 lb/10^6 gal
17.12 mg/l
17.118 ppm, based on density of 1 g/cm^3

grain/gallon (Imperial) =

14.25 g/m^3
14.255 ppm

grain/milliliter: GRAIN/CUBIC CENTIMETER

grain (apoth): grain (apothecary)

grain (apoth) =

0.0166 dr	0.002083 oz
0.0648 g	0.05 scruple

grain (avdp); grain (avoirdupois): GRAIN
grain (troy) =

0.324 carat	0.002083 oz
0.0166 dr	0.04167 pwt
0.0648 g	

grain alcohol: ethanol, ethyl alcohol
gram: metric unit of weight established as the weight of 1 cm³ of water
  at 39.2°F.

gram =

100 cg	1000 mg
0.1 dag	1 E−04 myriagram
10 dg	0.009807 N
0.2572 dr (apoth/troy)	0.03215 oz (apoth/troy)
0.5644 dr (avdp)	0.03527 oz (avdp)
980.665 dyn	0.0709 pdl
15.43 gr	1 pond
0.01 hg	0.643 pwt
9.807 E−05 J/cm	1 E−05 quintal
0.009807 J/m	0.77162 scruple (apoth)
0.001 kg	1.10231 ton
0.002679 lb (apoth/troy)	9.8426 E−07 ton (long)
0.002205 lb (avdp)	1 E−06 ton (metric)
1 E+06 μg	

gram (force) = 980.665 dyn
gram/centimeter =

980.665 dyn/cm	100 kg/km
2.54 g/in.	0.1 kg/m
39.1983 gr/in.	0.0672 lb/ft

gram/centimeter — *continued*

0.0056 lb/in.	0.180166 pdl/in.
2540 mg/in.	0.1 ton (metric)/km
0.98 07 N/m	

gram/centimeter/second: metric unit for absolute or dynamic viscosity.
gram/centimeter/second: POISE
gram/cubic centimeter =

980.665 dyn/cm^3	9.71 lb/gal (dry)
1 g/ml	8.3454 lb/gal (liquid)
15.433 gr/cm^3	0.03613 lb/in.^{3}
100 kg/hl	1685 lb/yd^3
1000 kg/m^3	1.1624 pdl/in.^{3}
3.405 E − 07 lb/cmil-ft	1.94 slug/ft^3
62.43 lb/ft^3	

gram/cubic foot = 0.02832 g/m^3
gram/cubic meter: PART/MILLION
gram/foot = 1.8372 E − 04 lb/in.
gram/inch =

0.3937 g/cm	0.002205 lb/in.
0.02645 lb/ft	1000 mg/in.

gram/liter =

58.417 gr/gal	3.612 E − 05 lb/in.^{3}
0.06243 lb/ft^3	1000 mg/l
0.008345 lb/gal (liquid)	1000 ppm, based on density of 1 g/cm^3

gram/milliliter: GRAM/CUBIC CENTIMETER
gram/square centimeter =

9.678 E − 04 atm	0.028959 in. (mercury, 0°C, 32°F)
9.8066 E − 04 bar	0.3937 in. (water, 4°C, 39.2°F)
980.66 baryes	0.001 kg/cm^2
0.073556 cm (mercury, 0°C, 32°F)	10 kg/m^2
980.665 dyn/cm^2	0.098066 kPa
0.03281 ft (water, 4°C, 39.2°F)	2.0482 lb/ft^2

gram/square centimeter — *continued*

0.01422 lb/in.^{2}	0.22757 oz/in.^{2}
0.98066 mbar	98.066 Pa
735.56 μm (mercury, 0°C, 32°F)	0.4576 pdl/in.^{2}
0.7356 mm (mercury, 0°C, 32°F)	0.001024 ton/ft^2
98.066 N/m^2	0.7356 torr

gram-calorie; calorie; calorie-gram: metric unit of heat energy required to raise the temperature of 1 g of pure water 1°C, specifically from 14.5 to 15.5°.

gram-calorie =

0.003968 Btu	1.58 E −06 hp-h (metric)
41.293 cm^3-atm	4.184 J
4.184 E +07 erg	0.001 kg-cal
0.0014582 ft^3-atm	0.4267 kg-m
3.086 ft-lb	1.1624 E −06 kWh
99.287 ft-pdl	0.04129 l-atm
4.2665 E +04 g-cm	0.0011624 Wh
1.559 E −06 hp-h	4.1846 W-s

gram-calorie/gram =

1.8 Btu/lb	4184 J/kg

gram-calorie/gram/°C = Btu/lb/°F
gram-calorie/gram/K = Btu/lb/°R
gram-calorie/hour =

0.003968 Btu/h	1.1623 E −06 kW
1.1623 E −04 erg/s	0.0011623 W
1.5586 E −06 hp	

gram-calorie/minute =

0.003968 Btu/min	7.11 E −06 hp (boiler)
6.9737 E +05 erg/s	0.069737 J/s
185.16 ft-lb/h	6.9737 E −05 kW
9.352 E −05 hp	0.069737 W

gram-calorie/mole; calorie/mole = g-cal/g molecular weight
gram-calorie/second =

14.286 Btu/h	0.005681 hp (metric)
0.003968 Btu/s	0.001 kg-cal/s
4.184 E + 07 erg/s	0.004184 kW
185.16 ft-lb/min	0.7356 torr
3.086 ft-lb/s	4.184 W
0.005611 hp	

gram-calorie/second/square centimeter =

1.32721 E + 04 Btu/h/ft²
4.184 W/cm²

gram-calorie/second-square centimeter-°C =

7373.38 Btu/h-ft²-°F
4.1840 E + 04 W/m²/K

gram-calorie/square centimeter =

3.6867 Btu/ft²
4.1840 E + 04 J/m²

gram-calorie-centimeter/hour-square centimeter-°C =

0.0671969 Btu-ft/h-ft²-°F	41.8 kWh-cm/h-m²-°C
0.806363 Btu-in./h-ft²-°F	0.851 kWh-in./h-ft²-°F
3.6 E + 04 kg-cal-cm/h-m²-°C	

gram-calorie-centimeter/second-square centimeter-°C =

2903 Btu-in./h-ft²-°F	850 J-in./s-ft²-°F
4.184 J-cm/s-cm²-°C	418.4 W-m/m²/K

gram-calorie-second = 6.31531 E + 33 Planck's constant
gram-centimeter =

9.301 E − 08 Btu	980.665 erg
980.665 dyn-cm	7.233 E − 05 ft-lb

gram-centimeter — *continued*

0.002327 ft-pdl	1 E − 05 kg-m
2.3438 E − 05 g-cal	2.7241 E − 11 kWh
3.65 E − 11 hp-h	9.8067 E − 05 N-m
9.807 E − 05 J	2.72407 E − 08 Wh
2.3438 E − 08 kg-cal	9.8067 E − 05 W-s

gram-centimeter/second/second = 1 dyn (force)
gram-centimeter-second =

9.301 E − 08 Btu/s	1.3151 E − 07 hp
980.665 erg/s	9.8067 E − 05 J/s
7.233 E − 05 ft-lb/s	9.8067 E − 08 kW
2.344 E − 05 g-cal/s	9.8067 E − 05 W

gram equivalent weight; equivalent weight: mass of an element equal to its atomic weight divided by its valence.

gram molecular weight: mol, mole, where the weight of a material, in grams, is numerically equal to its molecular weight.

gram-second/square centimeter: metric unit of absolute (dynamic) viscosity of a fluid.

gram-second/square centimeter = 980.665 P

GRAVITATIONAL ACCELERATION: the continuous increase of velocity of a free-falling body due to its own weight, where

E = kinetic energy, ft-lb
g = gravitational acceleration constant, 32.174 ft/s^2
S = distance in space that body falls, ft
t = time, s
V = velocity, ft/s
W = weight of body, lb

🔟 distance body falls in t seconds,

$$S = 0.5t^2 g = 16.087t^2$$

🔟 kinetic energy of falling body,

$$E = WS$$

GRAVITATIONAL ACCELERATION — *continued*

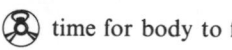 time for body to fall S feet,

$$t = 0.25S^{1/2}$$

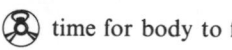 velocity of falling body, in t seconds,

$$V = gt = 32.174t = 2S/t = 8.02S^2$$

gravitational acceleration constant:

$$g = 980.24 \text{ cm/s}^2$$
$$= 32.174 \text{ ft/s}^2$$
$$= 9.8024 \text{ m/s}^2$$

gravitational system of units: English system using the units of force in pounds and mass in slugs.

gross =

12 dozen
144 items

Gunter's chain: CHAIN, GUNTER
Gunter's link: LINK, GUNTER

H

H: hydrogen (element)
Hagen–Poiseuille law: FLUID FLOW
hand: unit of measure used for measuring the height of a horse.
hand =

10.16 cm
4 in.

hardness of materials: BRINELL HARDNESS NUMBER, ROCKWELL HARDNESS
NUMBER
He: helium (element)
heat capacity: HEAT TRANSFER
heat of combustion: HEAT VALUE OF A MATERIAL
heat of fusion: HEAT TRANSFER
heat of solidification: HEAT TRANSFER
heat of sublimation: HEAT TRANSFER
heat of vaporization: HEAT TRANSFER
HEAT TRANSFER: where

a = coefficient of linear thermal expansion,
length change/unit length/°F
A = area of contact surface, ft^2
b = coefficient of cubical expansion, volume change/unit volume/°F
B = heat energy absorbed, Btu
C = specific heat, Btu/lb/°F
C_P = specific heat at constant pressure, Btu/lb/°F

HEAT TRANSFER — *continued*

C_V = specific heat at constant volume, Btu/lb/°F
d = depth (thickness), in.
e = radiation factor
E = electrical potential, volts
F = radiation factor, combined for two surfaces
h = convection coefficient, Btu/hr-ft²-°F
h_F = heat of fusion, Btu/lb
h_V = heat of vaporization, Btu/lb
H = heat flow, Btu/hr
I = electric current flow, amperes
k = conduction coefficient, Btu-in./hr-ft²-°F
L = length, in.
Q = flow, ft³/min
r = resistance, electrical, ohms
R = resistance, thermal, hr-ft²-°F/Btu
S = Stefan–Boltzmann constant, 0.174 E – 08 Btu/hr-ft²-°R⁴
t_1 = initial temperature, °F
t_2 = final temperature, °F
t_H = higher temperature, °F
t_L = lower temperature, °F
t_M = melting temperature of material, °F
T = absolute temperature, °R
T_H = absolute temperature of emitting surface, °R
T_L = absolute temperature of absorbing surface, °R
U = overall coefficient, Btu/hr-ft²-°F
V = volume, in.³
w = weight, lb
W = electric power, watts

absorptivity: radiation factor   *
black body: radiation, black body   *
coefficient of cubical expansion: thermal expansion, cubical coefficient   *
coefficient of cubical expansion, thermal: thermal expansion, cubical coefficient   *
coefficient of heat transfer, conduction: conduction coefficient   *
coefficient of heat transfer, convection: convection coefficient   *
coefficient of heat transfer, overall: overall coefficient   *

HEAT TRANSFER — *continued*

coefficient of linear expansion, thermal: thermal expansion, linear coefficient *

coefficient of thermal conductivity: conduction coefficient *

conductance: measure of heat flow for a given thickness of a material and difference in temperature from outside to outside, Btu-in./h-ft²-°F.

conduction: heating process whereby heat energy is transferred from a body of higher temperature to a body of a lower temperature by direct contact of the two bodies, or within one body when the heat flows from a higher-temperature zone to a lower-temperature zone.

 heat by conduction,

$$H = kA(t_H - t_L)/d$$

conduction coefficient of heat transfer (*k*): coefficient of thermal conductivity; the heat flow rate in Btu/hr per ft² (of surface area) per °F (temperature difference) for each inch of material thickness. Transfer of heat is affected by the difference in temperature on each side of the object and the specific material, Btu-in./hr-ft²-°F.

conduction coefficient (*k*) for various materials (Btu-in./hr-ft²-°F):

acetic acid: 1.188

acetone: 1.116

air: 0.177

aluminum: 1420

aluminum, cast: 1000

ammonia, liquid: 2.448

      vapor: 0.160

antinomy: 128

argon: 0.113

asbestos: 1.0

asbestos board: 0.48

asbestos paper: 0.5

asphalt: 4.8

benzene, liquid: 0.996

      vapor: 0.072

beryllium: 1014

bismuth: 51

boiler scale: 13

brass, 750

brick, alumina: 5.8

   common: 5

   fire: 9.0

   hard: 9.2

   insulating: 1.0

   magnesia: 20.6

   silica: 5.8

brine: 1.13

bronze: 486

cadmium: 640

cambric, varnished: 1.74

carbon: 165

carbon dioxide: 0.107

carbon disulfide: 1.112

carbon monoxide: 0.145

HEAT TRANSFER — *continued*

conduction coefficient (*k*) for various materials — *continued*

carbon tetrachloride, liquid: 72
     vapor: 0.06
carborundum: 1.45
cardboard: 1.45
castor oil: 1.248
cement, dry: 6.25
    wet: 0.49
chalk: 0.81
charcoal: 0.64
chromium: 465
cinders: 0.845
coal: 0.87
cobalt: 479
coke: 1.27
columbium: 378
concrete, dry: 6.0
    wet: 8.4
constantan: 150
copper: 2660
copper, annealed: 2715
cork: 0.30
corkboard: 0.28
cotton: 0.125
cotton batting, loose: 0.32
    packed: 0.21
delrin: 1.6
earth: 11.6
ethane vapor: 0.144
ethyl alcohol, liquid: 1.26
    vapor: 0.1068
ethylene glycol: 1.788
ethylene vapor: 0.130
felt: 0.29
fiberboard: 0.33
fiberglass: 1.44
fiber, red: 3.19
freon F12: 0.492
freon F22: 0.6
fuel oil: 1.02

gasoline liquid: 1.0
germanium: 410
german silver: 232
glass: 6
glass, block: 0.42
    plate: 7.42
glasswool: 0.33
glycerine liquid: 2.0
gold: 2028
granite: 13
graphite: 35
gravel: 2.58
gypsum: 3.0
helium: 1.02
hydrogen: 1.248
ice: 15.6
inconel: 104
indium: 167
invar: 76
iridium: 408
iron, cast: 316
    pure: 466
    wrought: 417
kapok: 0.238
kerosene, liquid: 1.05
lampblack: 0.203
lead: 240
leather: 1.22
limestone: 4.8
linoleum: 1.32
magnesia: 0.87
magnesium: 1100
manganin: 435
marble, hard: 24
    soft: 4
masonite: 0.33
mercury: 58
methane: 0.21
methyl alcohol: 1.404

HEAT TRANSFER — *continued*

conduction coefficient (*k*) for various materials — *continued*

mica: 2.49
mineral wool: 0.28
molybdenum: 1014
monel: 208
nichrome: 93
nickel: 625
niobium: 378
nitrogen: 0.1788
nylon: 1.7
oil, lubricating: 1.15
oxygen: 0.1788
palladium: 486
paper: 0.90
paraffin: 1.60
pasteboard: 1.30
perlite: 2.28
petroleum: 1.03
phenol: 1.32
phosphor bronze: 750
plaster: 3.77
plasterboard: 3.04
plaster of paris: 1.22
platinum: 482
polystyrene: 1.2
porcelain: 12.5
pumice: 1.24
rhodium: 600
rockwool: 0.30
rubber, hard: 1.25
    soft: 7.7
sand, dry: 2.4
    wet: 7.2
sandstone: 16
sawdust: 0.40
silica, fused: 7.39
silicon: 580
silver: 2904
slate: 13.9
snow: 1.74

steam at 212°F: 0.156
steel, carbon: 333
    stainless: 115
    structural: 400
stone, solid: 12.5
stucco: 12.0
tantalum: 320
teflon: 1.7
terra cotta: 6.67
terrazzo: 12.0
thorium: 257
tile: 12.0
tin: 450
titanium: 285
toluene: 0.924
tungsten: 1160
turpentine: 0.84
uranium: 188
urethane, foamed: 0.84
vermiculite: 0.52
water: 4.2
wood, ash: 1.08
    balsa: 0.33
    birch: 1.01
    cedar: 0.654
    cypress: 0.666
    fir: 0.624
    hemlock: 0.80
    mahogany: 0.916
    maple: 1.125
    oak: 1.0
    pine, white: 0.80
        yellow: 0.9
    redwood: 0.804
    spruce: 0.696
wool, mineral: 0.32
    pure: 0.40
zinc: 770
zirconium: 146

HEAT TRANSFER — *continued*

convection: heating process whereby heat is transferred by the motion of a heated medium (gas or liquid) over the body to be heated. Free convection exists where motion is due to gravity. Forced convection is where the heated medium is moved with blowers or pumps.

heat by convection,

$$\circledS \; H = hA(t_H - t_L)$$

convection coefficient of heat transfer (h): film coefficient of heat transfer, unit surface conductance. In convection heating, the heat flow rate in Btu/hr per ft^2 (of surface area per °F (temperature difference). Transfer of heat is affected by the difference in temperatures of the air (or fluid) movement across a surface. Coefficients of heat transfer are listed in ASHVE guide and other heating handbooks. Units are Btu/hr-ft^2-°F.

cubical expansion, thermal: thermal expansion, cubical   *

dielectric heating: material to be heated absorbs heat when high-frequency, high-voltage electrical power is applied across the material.

electric heat: heat energy is generated in a material (conductors) by the flow of electrical current through the material.

$$H = 3.414 I^2 r$$
$$W = I^2 r = EI$$

emissivity: radiation factor   *

film coefficient of heat transfer: convection coefficient   *

fusion heat: heat of fusion   *

heat capacity: specific heat   *

heat energy required:

to heat air,

$$H = 1.08 Q(t_2 - t_1)$$

to heat a material,

$$B = wC(t_2 - t_1)$$

to melt a material,

$$B = wC(t_M - t_1)$$

HEAT TRANSFER — *continued*

heat of combustion: HEAT VALUE OF A MATERIAL
heat of evaporization: heat of vaporization   *
heat of fusion: heat of solidification; quantity of heat energy absorbed by the changing of a material from a solid to a liquid (or from a liquid to a solid). A form of latent heat.

$$B = wh_F$$

heat of fusion (h_F) of various materials (Btu/lb):

acetic acid: 80
acetone: 42
aluminum: 172
ammonia: 150
antimony: 70
argon: 12
asphalt: 40
babitt, lead type: 26
    tin type: 34
beeswax: 76
benzene: 56
benzoic acid: 61
benzol: 56
beryllium: 570
bismuth: 23
brass, red: 87
    yellow: 71
bromine: 29
bronze/aluminum: 99
cadmium: 24
camphor: 19
carbon disulfide: 24.8
carbon tetrachloride: 74.8
cesium: 7
chlorine: 41
chloroform: 33.14
cobalt: 115
copper: 75
ether: 41.4
ethyl alcohol: 46

ethylene glycol: 77.9
freon F12: 14.8
    F22: 78.7
german silver: 86
glycerine: 76.5
gold: 29
hydrogen: 27
hydrogen peroxide: 158
ice: 144
iodine: 28
iron, cast: 40
    pure: 65
lead: 10
lithium: 286
magnesium: 130
manganese: 115
mercury: 5
methyl alcohol: 30
molybdenum: 126
monel: 117
nickel: 131
nitrogen: 11
octane: 78
oxygen: 6
palladium: 65
paraffin: 63
phenol: 45
phosphorus: 9
platinum: 49
potassium: 26

HEAT TRANSFER — *continued*

heat of fusion (h_F) of various materials — *continued*

propane: 34.38
silicon: 607
silver: 45
slag: 90
sodium: 49
solder 50/50: 17
sulfur: 17

tellurium: 13
tin: 25
toluene: 30.9
tungsten: 79
water: 144
Wood's metal: 17
zinc: 46

heat of solidification: heat of fusion   ∗

heat of sublimation: quantity of heat flow resulting from the changing of a material directly from a solid to a vapor (or from a vapor to a solid). A form of latent heat.

heat value of a material: HEAT VALUE OF A MATERIAL

heat of vaporization: heat of evaporization; quantity of heat energy flow (at contant temperature) that results from the changing of a liquid to a vapor (or vapor to a liquid). A form of latent heat.

$$B = wh_V$$

heat of vaporization (h_V) of various materials (Btu/lb)

acetic acid: 174
acetone: 220
alcohol: 365
aluminum: 4680
ammonia: 565
antimony: 703
argon: 68
barium: 1120
benzene: 170
benzine: 166
benzol: 170
bismuth: 395
bromine: 82
butane: 166
cadmium: 409
carbon dioxide: 159
carbon disulfide: 152
carbon tetrachloride: 84

chlorine: 121
chloroform: 106
chrome refractory: 105
ethane: 175
ether: 160
ethyl acetate: 180
ethyl alcohol: 370
ethyl bromide: 109
ethyl chloride: 169
ethylene glycol: 344
freon F11: 84
    F12: 68
    F13: 64
    F22: 94
fuel oil: 147
gasoline: 140
glycerine: 420
gold: 732

HEAT TRANSFER — *continued*

heat of vaporization (h_v) of various materials — *continued*

heptane: 138
hexane: 150
hydrogen: 194
hydrogen peroxide: 655
iodine: 72
kerosene: 110
mercury: 117
methane: 248
methyl alcohol: 481
methyl chloride: 180
naphtha: 184
naphthalene: 136
nickel: 2670
nitric acid: 207
nitrogen: 87

octane: 156
oil, light: 147
oxygen: 92
pentane: 154
propane: 147
propyl alcohol: 290
sulfur: 652
sulfur dioxide: 164
toluene: 150
toluol: 155
trichloroethylene: 100
turpentine: 133
water: 970
xylene: 147
zinc: 758

induction heating process: electrically conductive material is heated when held in an induction heater consisting of electrical power applied to a copper coil that encloses the material, without direct contact. The material is subjected to an AC magnetic field, produced by the coil, which heats the material. Magnetic materials are heated by eddy currents and hysteresis; nonmagnetic materials are heated by eddy currents.

infrared heating: form of radiation heating, utilizing special electrical infrared lamps or gas-fired infrared heaters.

latent heat: transfer of heat resulting from the physical change of a substance (at a constant pressure and constant temperature) with no work applied, as from a solid to a liquid, from a liquid to a gas, from a gas to a liquid, or a liquid to a solid. Heat of fusion, heat of sublimination, and heat of vaporization are forms of latent heat.

linear coefficient of thermal expansion: thermal expansion, linear coefficient ∗

linear expansion, thermal: thermal expansion, linear ∗

overall coefficient of heat transfer (U): total heat transfer coefficient for heat transmitted by conduction, convection, and radiation through a combination of materials. Values of overall coefficient for com-

HEAT TRANSFER — *continued*

binations of various materials and thicknesses may be found in the ASHVE guide, heating handbooks, or by calculation. It is the reciprocal of thermal resistance. Units are Btu/hr-ft^2-°F.

$$U = \frac{1}{R}$$

$$\frac{1}{R} = \frac{1}{1/h_1 + d_1/k_1 + d_2/k_2 + \cdots + d_n/k_n + 1/h_2}$$

overall transfer of heat: heat transfer through a combination of materials, as through a wall.

$$H = U A (t_H - t_L)$$

Planck's law: all substances emit radiation depending on the temperature of the surface and its radiation wavelength.

radiant heating: radiation heating   *

radiation, black body: term for a material, not necessarily black in color, that is a perfect emitter or absorber of radiant energy

radiation constant (S): Stefan–Boltzmann constant,

$$S = 0.174 \text{ E–08 Btu/hr-ft}^2\text{-°R}^4$$

radiation emission rate: Stefan–Boltzmann law; the rate of emission of radiant energy from the surface of a body is proportional to the fourth power of the absolute temperature of the body.

$$H = 0.174 \text{ E}-08 \ eAT^4$$

radiation factor (e): absorptivity; emissivity; the ratio of the rate of radiant energy emitted (or absorbed) by the surface of a material to the rate emitted (or absorbed) by a black body. Radiation factor varies at different temperatures.

radiation factor combined for two surfaces (F):

$$F = 1/e_1 + 1/e_2 - 1$$

HEAT TRANSFER — *continued*

radiation factor (*e*) for various materials:

aluminum, dull: 0.22
        oxidized: 0.85
        polished: 0.04
asbestos board: 0.95
asbestos slate: 0.96
asphalt: 0.94
BLACK BODY: 1.00
brass, dull: 0.25
      oxidized: 0.61
      polished: 0.03
brick, light color: 0.90
      red: 0.92
      yellow: 0.90
carbon: 0.81
ceramic tile: 0.93
chromium: 0.03
concrete: 0.90
copper, dull: 0.25
        oxidized: 0.80
        polished: 0.04
enamel: 0.92
fire brick: 0.90
glass: 0.92
gold, polished: 0.02
ice: 0.97
iron, cast: 0.83
      oxidized: 0.82
      polished: 0.25
      rusted: 0.90
      wrought: 0.20
lacquer: 0.92
lead, dull: 0.28
      polished: 0.07

magnesium: 0.07
marble: 0.95
mercury: 0.12
monel: 0.03
nickel, oxidized: 0.39
      polished: 0.04
oil: 0.80
paint, aluminum: 0.5
      bronze: 0.5
      dark: 0.90
      red/green: 0.90
paper: 0.86
plaster: 0.91
platinum: 0.04
porcelain: 0.92
quartz: 0.93
roofing paper: 0.92
rubber: 0.91
sand: 0.95
sawdust: 0.92
silver, polished: 0.025
slate: 0.94
steel, galvanized: 0.25
      oxidized: 0.85
      polished: 0.07
stone: 0.90
tile, red: 0.90
tin, clean: 0.08
      polished: 0.03
tungsten: 0.03
water: 0.96
wood: 0.87
zinc: 0.10

radiation heating: radiant heating, process whereby heat is emitted from a higher-temperature surface into space and absorbed by a lower-temperature surface through the natural radiation of heat waves.

HEAT TRANSFER — *continued*

⊗ radiation heat loss:

> heat emission rate of a surface,

$$H = 0.174 \text{ E}-08 \ eAT^4$$

radiation heat transfer:

> between two parallel plates,

$$H = 0.174 \text{ E}-08 \ eA(T_H^4 - T_L^4)$$

> between two parallel surfaces of different emissivity rates,

$$H = 0.174 \text{ E}-08 \ A(T_H^4 - T_L^4)/F$$

resistance, thermal: measure of resistance to heat flow, that is, the reciprocal of conductance.

$$R = 1/k$$

sensible heat: the heat of ambient air as indicated by the dry bulb temperature; also the heat that changes the temperature of a body.

specific heat; heat capacity: amount of heat energy required to raise the temperature of 1 lb of material 1°F,

$$C = B/w(t_2 - t_1)$$

specific heat, constant pressure (C_P): for gases and vapors, the ratio of enthalpy change of the gas to its temperature change when pressure is held constant. Units are $Btu/ft^3/°F$ or $Btu/lb/°F$.

specific heat for gases and vapors at constant pressure (C_P) (Btu/ lb/°F):

acetic acid: 0.412	carbon dioxide: 0.202
acetylene: 0.383	carbonic acid: 0.217
air at 70°F: 0.238	carbon monoxide: 0.243
at 200°F: 0.242	chlorine: 0.121
at 1000°F: 0.25	chloroform: 0.147
alcohol: 0.453	ethane: 0.409
ammonia: 0.508	ethylene: 0.40
argon: 0.123	fluorine: 0.32
butane: 0.397	freon F12: 0.15

HEAT TRANSFER — *continued*

specific heat for gases and vapors at constant pressure (C_p) — *continued*

helium: 1.25
hydrochloric acid: 0.195
hydrogen: 3.42
hydrogen sulfide: 0.245
isobutane: 0.398
methane: 0.532
neon: 0.443

nitrogen: 0.244
nitrous oxide: 0.226
oxygen: 0.218
propane: 0.576
steam at 212°F: 0.48
sulfur dioxide: 0.154
water vapor: 0.45

specific heat, constant volume (C_v): for gases and vapors, the ratio of internal energy changes of the gas to its temperature change when volume is held constant. Units are Btu/lb/°F.

specific heat for gases and vapors at constant volume (C_v) (Btu/ lb/°F):

acetylene: 0.303
air at 70°F: 0.171
alcohol: 0.399
ammonia: 0.399
argon: 0.074
butane: 0.32
carbon dioxide: 0.155
carbonic acid: 0.171
carbon monoxide: 0.177
chlorine: 0.09
ethane: 0.342

ethylene: 0.33
freon F12: 0.13
helium: 0.75
hydrogen: 2.43
isobutane: 0.358
methane: 0.403
nitrogen: 0.174
oxygen: 0.156
propane: 0.353
steam at 212°F: 0.36
sulfur dioxide: 0.123

specific heat (C) for various materials (Btu/lb/°F)

acetic acid: 0.472
acetone: 0.544
alcohol: 0.65
alumina: 0.20
aluminum: 0.225
aluminum foil: 0.24
aluminum oxide: 0.183
ammonia: 1.08
ammonium sulfate: 0.283
andalusite: 0.228

aniline: 0.514
antimony: 0.050
arsenic: 0.0822
asbestos: 0.20
ashes: 0.20
asphalt: 0.40
babbitt, lead types: 0.039
 tin type: 0.071
baking soda: 0.231
barium: 0.068

HEAT TRANSFER — *continued*

specific heat (*C*) for various materials — *continued*

beeswax: 0.82
benzene: 0.414
benzine: 0.423
benzoic acid: 0.287
beryllium: 0.52
bismuth: 0.031
borax: 0.385
boron: 0.307
brass, red: 0.104
      yellow: 0.105
brick, common: 0.22
      furnace: 0.20
      hard: 0.24
bromine: 0.107
bronze, aluminum: 0.126
      copper/tin: 0.09
      tobin: 0.107
cadmium: 0.057
calcium: 0.170
calcium carbonate: 0.210
calcium chloride: 0.676
camphor: 0.44
carbon: 0.204
carbon tetrachloride: 0.215
castor oil: 0.434
cellulose: 0.32
cement, Portland: 0.271
cerium: 0.0448
cesium: 0.0482
chalk: 0.215
charcoal: 0.20
chloroform: 0.25
chrome, refractory: 0.20
chromium: 0.12
cinders: 0.18
clay: 0.222
coal: 0.30
cobalt: 0.108
coke: 0.28

columbium: 0.065
concrete: 0.16
constantan: 0.098
copper: 0.095
copper sulfate: 0.848
cork: 0.419
cork, granulated: 0.43
corkboard: 0.204
corundum: 0.198
cotton: 0.362
cottonseed oil: 0.47
cream: 0.780
cupric oxide: 0.227
cuprous oxide: 0.111
diamond: 0.12
dolomite: 0.222
earth: 0.44
ebonite: 0.33
ether: 0.503
ethyl acetate: 0.478
ethyl alcohol: 0.58
ethyl bromide: 0.21
ethyl ether: 0.529
ethyl iodide: 0.25
ethylene glycol: 0.602
fiberglass board: 0.236
fire blanket: 0.5
fireboard: 0.5
fire brick: 0.26
freon F12: 0.232
      F22: 0.300
fuel oil: 0.50
galena: 0.0466
gallium: 0.080
gasoline: 0.53
germanium: 0.0737
german silver: 0.0945
glass: 0.20
glass block: 0.170

HEAT TRANSFER — *continued*

specific heat (*C*) for various materials — *continued*

glass wool: 0.210
glue, $\frac{2}{3}$ water: 0.895
glycerine: 0.576
gold: 0.033
granite: 0.192
graphite: 0.201
gypsum: 0.259
hematite: 0.1645
heptane: 0.55
hexane: 0.60
humus soil: 0.44
ice: 0.52
indium: 0.057
insulating brick: 0.20
iodine: 0.054
iridium: 0.032
iron, cast: 0.13
    pure: 0.11
    wrought: 0.115
kaolin: 0.22
kapok fiber: 0.32
kerosene: 0.50
lanthanum: 0.0448
lava: 0.197
lead: 0.032
lead, melted: 0.04
lead oxide: 0.049
leather: 0.36
limestone: 0.216
linseed oil: 0.44
lithium: 1.04
lodestone: 0.156
magnesia: 0.222
magnesite refractory: 0.27
magnesium: 0.25
magnesium oxide: 0.25
magnetite: 0.156
manganese: 0.12
manganin: 0.097

marble: 0.21
masonry brickwork: 0.22
masonry, rubble: 0.22
mercury, 0.033
methyl alcohol: 0.60
methyl chloride: 0.385
mica: 0.207
milk: 0.847
mineral wool: 0.20
molasses: 0.60
molybdenum: 0.065
monel: 0.129
naphtha: 0.493
nickel: 0.11
nitric acid: 0.445
octane: 0.52
oil: 0.5
oil, lubricating: 0.4
olive oil: 0.47
osmium: 0.0311
oxalic acid: 0.416
palladium: 0.0714
paper: 0.45
paraffin: 0.62
petroleum: 0.51
phenol: 0.34
phosphorus: 0.183
pipe insulation: 0.20
pitch: 0.45
plaster: 0.25
platinum: 0.032
porcelain: 0.26
potassium: 0.17
potassium hydroxide: 0.876
propane: 0.576
pyrex: 0.196
pyrite: 0.1291
quartz: 0.188
quicklime: 0.217

HEAT TRANSFER — *continued*

specific heat (C) for various materials — *continued*

rhodium: 0.058
rock salt: 0.219
rockwool: 0.225
rosin: 0.525
rubber: 0.40
rubidium: 0.0802
ruthenium: 0.0611
sand: 0.195
seawater: 0.938
selenium: 0.068
shellac: 0.40
silica refractory: 0.23
silicon: 0.18
silicon carbide: 0.23
silver: 0.056
slag: 0.17
sodium: 0.253
sodium carbonate: 0.306
sodium chloride: 0.219
sodium hydroxide: 0.942
sodium nitrate: 0.23
solder 50/50: 0.04
steel, carbon: 0.116
　　　stainless: 0.12
　　　structural: 0.11
stone: 0.20
sugar: 0.30
sulfur, liquid: 0.234
　　　solid: 0.18
sulfuric acid: 0.336
talc: 0.209

tantalum: 0.04
tartaric acid: 0.287
tellurium: 0.048
thallium: 0.033
thorium: 0.0276
tile, hollow: 0.015
tin: 0.055
tin, melted: 0.064
titanium: 0.1125
toluene: 0.40
toluol: 0.49
tungsten: 0.034
turpentine: 0.41
type metal: 0.039
uranium: 0.028
vanadium: 0.115
vermiculite: 0.22
water: 1.00
wood, fir: 0.65
　　　hemlock: 0.65
　　　oak: 0.45
　　　pine: 0.47
Woods metal: 0.041
wool: 0.39
xylene: 0.42
zinc: 0.092
zinc oxide: 0.125
zinc sulfate: 0.174
zircon: 0.132
zirconium: 0.066

Stefan–Boltzmann constant: radiation constant　*
Stefan–Boltzmann Law: radiation emission rate　*
sublimation heat: heat of sublimation　*
thermal conductance: conductance　*
thermal conductivity coefficient: conduction coefficient　*
thermal expansion, cubical: change in volume per unit volume at constant

HEAT TRANSFER — *continued*

pressure, based on the increase in temperature of the material and its cubical expansion coefficient. The value of the thermal expansion cubical coefficient for a material is approximately three times its thermal expansion linear coefficient.

$$V_2 - V_1 = bV_1(t_2 - t_1)$$

thermal expansion, cubic coefficient (b) for materials (volume change $E - 04$/unit volume/°F at constant pressure):

acetic acid: 6.0	gasoline: 6.0
acetone: 8.3	glycerine: 2.7
benzene: 6.9	mercury: 1.011
bromine: 6.5	methyl alcohol: 7.2
butane: 9.0	octane: 4.0
carbon disulfide: 6.3	petroleum: 5.0
carbon tetrachloride: 6.87	phenol: 5.0
ether: 9.8	propane: 9.0
ethyl alcohol: 6.2	toluol: 6.3
fuel oil: 4.0	turpentine: 5.41
gases: 20.33	water: 1.15

thermal expansion, linear: change in length per unit length, based on an increase in temperature of the material and its linear expansion coefficient.

$$\bigotimes L_2 - L_1 = aL_1(t_2 - t_1)$$

thermal expansion, linear coefficient (a) for materials (length change $E - 06$/unit length/°F):

aluminum: 13.33	carbon, diamond: 0.67
aluminum, wrought: 12.8	cement: 5.9
antimony: 6.7	chromium: 3.4
beryllium: 6.9	cobalt: 6.7
bismuth: 7.5	columbium: 3.8
brass: 10.4	concrete: 7.0
brick: 3.1	constantan: 8.3
bronze: 10.0	copper: 9.3
cadmium: 16.6	delrin: 30.5
calcium: 13.9	ebonite: 42.8

HEAT TRANSFER — *continued*

thermal expansion, linear coefficient (*a*) for materials — *continued*

flint: 4.4
germanium: 3.2
german silver: 10.2
glass: 5.0
gold: 7.9
granite: 4.6
graphite: 4.4
hafnium: 3.7
ice: 28
inconel: 6.4
indium: 18.3
invar: 0.5
iridium: 3.61
iron, cast: 5.9
    malleable: 6.6
    pure: 6.61
    wrought: 6.7
lead: 16.1
limestone: 4.4
magnesium: 14.5
manganese: 12.9
marble: 5.6
masonry brickwork: 3.5
mica: 4.2
molybdenum: 2.7
monel: 7.8
nichrome: 7.2
nickel: 7.0
niobium: 4
nylon: 50
osmium: 3.7
palladium: 6.6
paraffin: 72
phenolic, molded: 20.8
phosphor bronze: 9.9
phosphorus: 69
plaster: 9.2
platinum: 4.9

polyethylene: 94.4
polystyrene: 38.9
porcelain: 2.0
quartz: 0.3
rhodium: 4.4
rubber: 44.4
ruthenium: 5.1
sandstone: 6.0
selenium: 20.4
silicon: 4.1
silver: 10.5
slate: 5.8
sodium: 34.0
solder: 13.9
steel, carbon: 6.5
    cast: 7.56
    stainless: 9.2
    structural: 7.0
stone: 6.0
sulfur: 35.6
tantalum: 3.6
teflon: 69
tellurium: 9.5
thorium: 6.9
tin: 14
titanium: 4.7
tungsten: 2.5
uranium: 3.9
vanadium: 4.3
wood, ash: 5.28
    beech: 1.43
    fir: 2.1
    maple: 3.5
    oak: 2.7
    pine: 3.0
    walnut: 3.66
zinc: 16
zirconium: 3.2

HEAT TRANSFER — *continued*

unit surface conductance: convection coefficient of heat transfer   *

vaporization heat: heat of vaporization   *

---

heat value of a material: heat of combustion; total heat energy released by a specific material when the material is completely burned.

heat value of various gases (Btu/ft^3):

acetylene: 1482
blast furnace gas: 100
butane: 3230
butylene: 3190
carbon monoxide: 323
coal gas: 650
coke oven gas: 638
ethane: 1766

ether: 4297
ethylene: 1575
gas, natural: 1040
hydrogen: 323
hydrogen sulfide: 638
methane: 1000
producers gas: 150
propane: 2515

heat value of various liquids (Btu/gal):

ammonia: 72,000
benzene, benzol: 137,000
benzine: 102,000
butane: 102,000
carbon disulfide: 61,000
crude oil: 142,000
diesel oil: 150,000
ethyl alcohol: 83,700
fuel oil no. 1: 135,740
    no. 2: 141,000

fuel oil no. 3: 141,350
    no. 4: 144,770
    no. 5: 149,000
    no. 6: 154,000
gasoline: 120,000
kerosene: 132,660
methyl alcohol: 64,400
oil: 137,000
propane: 92,225
xylene: 139,000

heat value of various materials (Btu/lb):

acetylene: 21,600
ammonia: 9670
benzene, benzol: 18,500
benzine: 17,900
blast furnace gas: 1200

butane: 21,200
butylene: 21,600
carbon: 14,500
carbon dioxide: 14,500
carbon disulfide: 5800

heat value of various materials — *continued*

carbon monoxide: 4370
charcoal: 13,000
coal: 14,000
coke: 13,000
coke oven gas: 19,000
crude oil: 19,000
diesel oil: 20,000
ethane: 22,300
ether: 22,000
ethyl alcohol: 12,400
ethylene: 21,320
fuel oil no. 1: 19,730
   no. 2: 19,600
   no. 3: 19,470
   no. 4: 19,200
   no. 5: 19,000
   no. 6: 18,750
gasoline: 20,000
hexane: 20,800
hydrogen: 61,000

hydrogen sulfide: 7100
kerosene: 19,800
methane: 23,700
methyl alcohol: 9600
naphthalene: 17,300
natural gas: 22,800
octane: 20,600
oil: 18,600
peat: 7600
pentane: 21,100
producer gas: 2400
propane: 21,700
propylene: 21,500
sawdust: 4500
straw: 5800
sulfur: 4020
toluene: 18,300
wood, yellow pine: 6500
   oak: 5500
xylene: 18,600

hectare =

2.471 acres
100 ares
1 E + 04 centares
1 E + 08 cm^2
1.0764 E + 05 ft^2

0.01 km^2
1 E + 04 m^2
0.003861 mi^2
395.37 square rods
1.19599 E + 04 yd^2

hecto; hekto: prefix, equals 100

hectogram =

10 dag
100 g
0.1 kg

0.268 lb (apoth/troy)
0.2205 lb (avdp)
7.09316 pdl

hectoliter =

2.838 bushels (level)	100 l
1 E+05 cm^3	1 E+05 ml
3.5316 ft^3	3381.5 oz (liquid)
26.42 gal (liquid)	11.35136 pecks
0.1 kl	

hectometer =

1 E+04 cm	0.1 km
10 dam	100 m
1000 dm	19.884 rods
328.084 ft	109.36 yd

hectowatt = 100 W
hekto: HECTO
hemisphere =

0.5 sphere	6.283185 sr

henry: ELECTRICAL
henry =

1 E+09 abhenrys	1000 mH
1 E+06 μH	1.11265 E−12 stathenry

henry/meter = 7.9577 E+05 G/Oe
hertz (Hz): measure of frequency.
hertz = 1 c/s
hexagon: six equal-sided figure with six equal angles.
hexagon section: where

a, b, h, x, y = dimensions, in.
A = cross-sectional area, in.^{2}
cg = center of gravity
I_{AA}, I_{BB} = moment of inertia, in.^{4}
I_P = moment of inertia, polar, in.^{4}
k_{AA}, k_{BB} = radius of gyration, in.
Z_{AA}, Z_{BB} = section modulus, in.^{3}
Z_P = section modulus, polar, in.^{3}

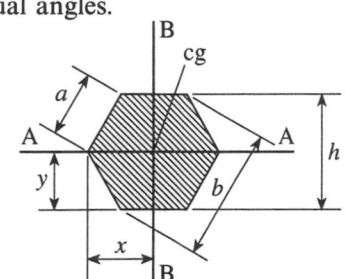

hexagon section — *continued*

area,

$$A = 2.598a^2 = 0.866h^2$$

center of gravity (cg), at center,

$$x = a = 0.5b = 0.57735h$$
$$y = 0.866a = 0.5h$$

distance across corners,

$$b = 2a = 1.1547h$$

distance across flats,

$$h = 1.732a$$

moment of inertia,

$$I_{AA} = I_{BB} = 0.54a^4 = 0.06h^4$$

moment of inertia, polar,

$$I_P = 0.1203h^4 = 1.0825a^4$$

radius of gyration,

$$k_{AA} = k_{BB} = 0.4564a = 0.2635h$$

section modulus,

$$Z_{AA} = 0.62a^3 = 0.12h^3$$
$$Z_{BB} = 0.54a^3 = 0.104h^3$$

section modulus, polar,

$$Z_P = 0.208h^3 = 1.0824a^3$$

Hf: hafnium (element)

Hg: mercury (element)

HHV: higher heating value

higher heating value (HHV): refers to the highest heating value of a fuel available when the condensed water vapors are removed from the products of combustion (gases).

Ho: holmium (element)
hogshead: type of barrel
hogshead =

8.422 ft^3	238.47 l
63 gal	0.2385 m^3
14,553 in.^{3}	

Hooke's law: STRENGTH OF MATERIALS
horsepower: term used for mechanical power as measurement of work done per unit time, originally established by British engineer James Watt as 33,000 ft-lb of work done by a horse in 1 min.
horsepower =

2546 Btu/h	0.076 hp (boiler)
42.433 Btu/min	1.014 hp (metric)
0.707 Btu/s	745.7 J/s
7.457 E+09 erg/s	641.6 kg-cal/h
1.98 E+06 ft-lb/h	10.693 kg-cal/min
33,000 ft-lb/min	0.178 kg-cal/s
550 ft-lb/s	76.04 kg-m/s
6.41582 E+05 g-cal/h	0.7457 kW
1.0693 E+04 g-cal/min	0.212 ton (refrigeration)
178 g-cal/s	745.7 W
7.604 E+06 g-cm/s	

horsepower (boiler): power required to evaporate 34.5 lb of water per hour at atmospheric pressure from 212°F to dry steam.
horsepower (boiler) =

33,492 Btu/h	1.4067 E+05 g-cal/min
558 Btu/min	2341 g-cal/s
9.3 Btu/s	1 E+08 g-cm/s
9.8095 E+10 erg/s	13.1548 hp
2.6046 E+07 ft-lb/h	13.337 hp (metric)
4.341 E+05 ft-lb/min	9809.5 J/s
7235 ft-lb/s	8440 kg-cal/h
8.44 E+06 g-cal/h	140.66 kg-cal/min

horsepower (boiler) — *continued*

2.34 kg-cal/s
1000 kg-m/s
9.8095 kW

2.789 ton (refrigeration)
9809.5 W

horsepower (brake): BRAKE HORSEPOWER
horsepower (metric) =

2511.31 Btu/h
41.853 Btu/min
0.69758 Btu/s
1 Cheval-Vapeur
7.35 E + 09 erg/s
32,548 ft-lb/min
542.475 ft-lb/s
176 g-cal/s
0.9863 hp

0.07498 hp (boiler)
633 kg-cal/h
0.546 kg-cal/min
0.17577 kg-cal/s
4500 kg-m/min
75 kg-m/s
0.7355 kW
735.448 W

horsepower/square foot = 42.433 Btu/ft^2/min
horsepower-hour =

2546 Btu
2.6845 E + 13 erg
935.6 ft^3-atm
1.98 E + 06 ft-lb
6.37 E + 07 ft-pdl
6.41582 E + 05 g-cal
2.7375 E + 10 g-cm
1.01387 hp-h (metric)

2.376 E + 07 in.-lb
2.6845 E + 06 J
641.6 kg-cal
2.737 E + 05 kg-m
0.7457 kWh
2.6493 E + 04 l-atm
745.7 Wh
2.6842 E + 06 W-s

horsepower-hour (metric) =

2511 Btu
1 Cheval-Vapeur hour
2.6487 E + 13 erg
1.9529 E + 06 ft-lb
6.327 E + 05 g-cal
0.98632 hp-h

2.6476 E + 06 J
633 kg-cal
2.7 E + 05 kg-m
0.73545 kWh
2.61298 E + 04 l-atm

horsepower-hour/pound =

  2546 Btu/lb   5918 J/g
  1.98 E+06 ft-lb/lb   0.001644 kWh/g
  1414 g-cal/g

horsepower-second = 550 ft-lb
hour =

  0.04167 d   3600 s
  60 min

humidity, absolute: in an air/water-vapor mixture, the ratio of weight of
  water vapor to weight of the mixture, where

  AH = absolute humidity
  W_M = weight of air/water-vapor mixture, lb/ft^3
  W_W = weight of water vapor, lb/ft^3

absolute humidity,

$$AH = W_W/W_M$$

humidity, relative: in an air/water-vapor mixture, the ratio of the weight
  of the air/water-vapor mixture at a specific dry bulb temperature
  to the weight of saturated water vapor at the same temperature.
  Relative humidity values for various dry and wet bulb temperatures of air may also be found in Psychrometric Charts, where

  RH = relative humidity
  W_M = weight of air/water-vapor mixture, lb/ft^3
  W_S = weight of saturated water vapor, lb/ft^3

relative humidity,

$$RH = W_M/W_S$$

relative humidity, percent,

$$RH\% = 100(RH)$$

humidity, specific: in an air/water-vapor mixture, the ratio of weight of water vapor per pound of dry air, where

> SH = specific humidity
> W_A = weight of dry air, lb/ft^3
> W_W = weight of water vapor, lb/ft^3

specific humidity,

$$SH = W_W/W_A$$

hundredweight: 100 lb

hydraulics; hydrokinetics: application of power obtained from liquids under pressure and in motion.

HYDRAULICS FORMULAS: where

> A = area of cylinder, in.^{2}
> C = cylinder fluid capacity, gal
> C_P = pump capacity, in.^{3}/rev
> d_C = inside diameter of cylinder, in.
> d_P = inside diameter of pipe, in.
> d_R = diameter of piston rod, in.
> d_S = inside diameter of sphere, in.
> D_C = outside diameter of cylinder, in.
> eff = pump efficiency, decimal
> F = force, lb
> h = head, ft (fluid)
> hp = horsepower
> N = speed, rpm
> P = pressure, lb/in.^{2}
> Q = fluid flow rate, gal/min
> s = stress, working (allowable), lb/in.^{2}
> S = cylinder stroke, in.
> t = wall thickness, in.
> T = torque, in.-lb
> V = velocity, ft/sec
> w = specific weight, lb/ft^3

cylinder area:

$$A = 0.7854d_C^2$$

HYDRAULICS FORMULAS — *continued*

cylinder fluid capacity:

piston end,

$$C = 0.0034Sd_C^2$$

rod end,

$$C = 0.0034S(d_C^2 - d_R^2)$$

cylinder fluid flow rate:

piston end,

$$Q = 2.448Vd_C^2$$

rod end,

$$Q = 2.448V(d_C^2 - d_R^2)$$

cylinder pressure, maximum (Lamé's formula):

$$P = s(D_C^2 - d_C^2)/(D_C^2 + d_C^2)$$

cylinder stress, allowable (Lamé's formula):

$$s = P(D_C^2 + d_C^2)/(D_C^2 - d_C^2)$$

cylinder velocity:

$$V = 0.3208Q/A$$

cylinder wall thickness, minimum:

$$t = 0.5d_C(s + P)^{1/2}/[(s - P)^{1/2} - 0.5d_C]$$

fluid pressure:

$$P = F/A$$

fluid velocity through piping:

$$V = 0.4085Q/d_P^2$$

force due to cylinder pressure:

$$F = PA$$

HYDRAULICS FORMULAS — *continued*

pipe stress due to fluid pressure:

$$s = 0.5 d_p P/t$$

pipe wall thickness required:

$$t = 0.5 d_p P/s$$

pressure due to fluid head:

$$P = 0.0069 wh$$

pressure head:

$$h = 144 P/w$$

pump capacity:

$$Q = C_P N/231$$

pump efficiency:

$$\text{eff} = \text{hp output/hp input}$$

pump efficiency, percent:

$$\text{eff\%} = 100(\text{eff})$$

pump horsepower input:

$$\text{hp} = QP/1714/\text{eff}$$

pump horsepower output:

$$\text{hp} = TN/63{,}025$$

pump speed:

$$N = 231 Q/C_P$$

pump torque:

$$T = 36.76 QP/N$$
$$= 63{,}025(\text{hp})/N$$

HYDRAULICS FORMULAS — *continued*

sphere wall thickness, minimum:

$$t = 0.25Pd_S/s$$

---

hydrogen atom mass = 1.6734 E − 24 g
hydrokinetics: HYDRAULICS
hydrostatics: study of liquid at rest.
hyperbolic logarithms: LOGARITHMS
hypotenuse: TRIANGLES, RIGHT
hysteresis coefficient: ELECTROMAGNETISM
hysteresis loss: ELECTROMAGNETISM

I

I: iodine (element)
ICE:

heat energy required to melt 1 kg/h:

0.0263 ton (refrigeration)

heat energy required to melt 1 kg/min:

4782 kg-cal/h
79.69 kg-cal/min

heat energy required to melt 1 lb/h:

143.44 Btu/h
2.39074 Btu/min
0.0119 ton (refrigeration)

latent heat: 144 Btu/lb
specific gravity: 0.913
specific weight: 57 lb/ft^3

ideal gas: GAS, IDEAL
IES: Illuminating Engineer's Society
ILLUMINATION: visible form of radiant energy; density of luminous flux falling on a surface measured in ft-cdl.

brightness: luminous intensity per unit area, reflected from a surface, measured in cd/in.^{2} or ft-L.

ILLUMINATION — *continued*

candela: candle   *

candle: candela, standard unit for measurement of candlepower of a light source. A light source of 1 cd produces 12.57 lm, also 1 cd equals 1 lm/steradian.

candle/square centimeter equivalent: CANDLE/SQUARE CENTIMETER

candle/square foot equivalent: CANDLE/SQUARE FOOT

candle/square inch equivalent: CANDLE/SQUARE INCH

candle/square meter equivalent: CANDLE/SQUARE METER

candlepower: luminous intensity; amount of light flow produced by a light source in a specific direction, measured in candles. One candlepower equals 12.57 lm.

foot-candle: amount of light on a surface; measure of illumination equivalent to the illumination produced by light flux of 1 lm/ft^2.

foot-candle equivalent: FOOT-CANDLE

foot-lambert: unit for measure of brightness of a surface emitting or reflecting light at a rate of 1 lm/ft^2. Foot-lambert is a function of foot-candles of illumination projected and the reflectivity of the surface. A perfect reflecting surface illuminated by 1 ft-cd has a brightness of 1 ft-L, however, because all surfaces absorb some light, brightness in foot-lamberts is equal to foot-candles of illumination multiplied by the percent of light reflected.

foot-lambert equivalent: FOOT-LAMBERT

lambert: metric unit for measure of brightness of a surface emitting or reflecting light at a rate of 1 lm/cm^2.

lambert equivalent: LAMBERT

lumen: total quantity of light emitted; measure of luminous flux as projected on 1 ft^2 area of a surface from a light source of 1 cdl a distance of 1 ft to each point on the surface area.

lumen equivalent: LUMEN

luminous flux: rate of flow of light; form of radiant energy producing brightness, measured in lumens.

luminous intensity: candlepower   *

lux: metric unit of illumination of a surface equivalent to 1 lm/m^2.

lux equivalent: LUX

photon: light quantum of radiant energy.

ILLUMINATION — *continued*

steradian: subtended solid angle of a sphere originating at the center of the sphere and extended to an area on the surface of the sphere; equal to the square of the sphere radius. Total solid angle around the point is equal to 12.57 steradians.

steradian equivalent: STERADIAN

---

impedance: ELECTRICAL
impulse: LINEAR MOTION
impulse, angular: ROTATION
In: indium (element)
inch =

2.54 E + 08 Å
0.001263 chain (Gunter)
8.33 E − 04 chain (Ramden)
2.54 cm
0.0556 cubit
0.00254 dam
0.254 dm
0.0222 ell
0.013889 fathom
0.0833 ft
1.263 E − 04 furlong
0.25 hand
2.54 E − 05 km

0.1263 link (Gunter)
0.0833 link (Ramden)
0.0254 m
2.54 E + 04 μm
1.578 E − 05 mi
1000 mils
25.4 mm
2.54 E + 07 nm
6.0225 picas (printer's)
2.54 E + 10 pm
72 points (printer's)
0.02778 yd

inch (mercury, 0°C, 32°F) =

0.03342 atm
0.03386 bar
2.54 cm (mercury)
3.3864 E + 04 dyn/cm^2
1.133 ft (water, 4°C, 39.2°F)
1.135 ft (water, 20°C, 68°F)
34.532 g/cm^2

13.5947 in. (water, 4°C, 39.2°F)
13.623 in. (water, 20°C, 68°F)
0.03453 kg/cm^2
345.3 kg/m^2
3.3864 kPa
70.726 lb/ft^2
0.49115 lb/in.^{2}

inch (mercury, 0°C, 32°F) — *continued*

0.0254 m (mercury)	3386.4 N/m^2
33.864 mbar	1131.62 oz/ft^2
2.54 E + 04 μm (mercury)	7.85847 oz/in.^{2}
25.4 mm (mercury)	3386.4 Pa
345.9 mm (water, 20°C, 68°F)	25.4 torr

inch (water, 4°C, 39.2°F) =

0.002458 atm	5.202 lb/ft^2
0.002491 bar	0.03613 lb/in.^{2}
0.18684 cm (mercury, 0°C, 32°F)	1.868 mm (mercurcy, 0°C, 32°F)
2490.8 dyn/cm^2	25.4 mm (water)
2.540 g/cm^2	249.1 N/m^2
0.07356 in. (mercury, 0°C, 32°F)	83.233 oz/ft^2
0.00254 kg/cm^2	0.578 oz/in.^{2}
25.3988 kg/m^2	249.1 Pa
0.2491 kPa	1.868 torr

inch (water, 20°C, 68°F) =

0.00245 atm	5.1936 lb/ft^2
0.002486 bar	0.03606 lb/in.^{2}
0.1865 cm (mercury, 0°C, 32°F)	1.865 mm (mercury, 0°C, 32°F)
2486 dyn/cm^2	25.4 mm (water)
2.535 g/cm^2	248.64 N/m^2
0.0734 in. (mercury, 0°C, 32°F)	0.5774 oz/in.^{2}
0.998 in. (water, 4°C, 39.2°F)	248.64 Pa
0.002535 kg/cm^2	1.865 torr
0.2486 kPa	

inch/hour =

2.54 cm/h	0.001667 in./min
0.0833 ft/h	2.7778 in./s
0.0013889 ft/min	1.578 E − 05 mi/h
2.3148 E − 05 ft/s	

inch/minute =

152.4 cm/h	0.0013889 ft/s
5 ft/h	0.016667 in./s
0.0833 ft/min	9.469 E − 04 mi/h

inch/second =

2.54 cm/s	60 in./min
300 ft/h	0.0254 m/s
5 ft/min	0.05682 mi/h
0.0833 ft/s	9.47 E − 04 mi/min
3600 in./h	

inch/second/second =

2.54 cm/s^2
0.0254 m/s^2

inch-ounce; ounce-inch =

0.0052 ft-lb
0.00706 N-m

inch-pound; pound-inch: unit of mechanical energy.
inch-pound =

1.0703 E − 04 Btu	1.152 kg-cm
1.1298 E + 06 dyn-cm	3.1374 E − 08 kWh
1.1298 E + 06 erg	0.11298 N-m
0.08333 ft-lb	3.1374 E − 05 Wh
4.2088 E − 08 hp-h	

inch-pound/inch = 4.4482 N-m/m
inductance: ELECTRICAL
induction heating: HEAT TRANSFER
inductive reactance: ELECTRICAL
inertia: a property of matter measured as the resistance of a body to change its velocity, whereby a force must be exerted on the body in order to accelerate it.

infrared heating: HEAT TRANSFER

infrared rays: portion of the electromagnetic spectrum with wavelengths ranging from 0.77 to 400 μm. Temperature intensities are inversely proportional to the wavelength, i.e., shorter wavelength, higher temperature.

International Electrical Units: ELECTRICAL

Ir: iridium (element)

isentropic process: THERMODYNAMICS

isobaric process: THERMODYNAMICS

isoceles triangles: TRIANGLES, ISOCELES

isochoric process: THERMODYNAMICS

isodynamic process: THERMODYNAMICS

isothermal process: THERMODYNAMICS

isovolume process: THERMODYNAMICS

J

joule, electrical energy: ELECTRICAL

joule, mechanical energy: metric unit for work done by a force of 1 N acting through a distance of 1 m.

joule =

9.4845 E − 04 Btu
9.869 cm³-atm
1 E + 07 dyn-cm
1 E + 07 erg
3.485 E − 04 ft³-atm
0.7376 ft-lb
23.73 ft-pdl
0.239 g-cal
1.019716 E + 04 g-cm
3.725 E − 07 hp-h

3.777 E − 07 hp-h (metric)
2.387 E − 04 kg-cal
0.10197 kg-m
0.001 kJ
2.778 E − 07 kWh
0.009869 l-atm
1 N-m
1 V-C
2.778 E − 04 Wh
1 W-s

joule/centimeter =

1 E + 07 dyn
1.0197 E + 04 g
100 J/m
10.197 kg

22.48 lb
100 N
723.3 pdl

joule/coulomb: VOLT

joule/cubic meter =

2.687 E − 05 Btu/ft³

3.591 E − 06 Btu/gal

joule/cubic meter/K $= 1.49$ E-05 Btu/ft^3/°F

joule/gram $=$

    0.43035 Btu/lb   2.778 E-07 kWh/g
    334.55 ft-lb/lb

joule/hour $=$

    1.69 E-04 hp-h/lb
    2.778 E-07 kW

joule/kilogram $= 4.303$ E-04 Btu/lb
joule/kilogram/K $= 2.39$ E-04 Btu/lb/°F
joule/meter: NEWTON
joule/radian $= 0.7375$ ft-lb
joule/second: WATT
joule-centimeter/second-square centimeter-°C $= 694$ Btu-in./h-ft^2-°F
joule-inch/second-square foot-°F $= 3.413$ Btu-in./h-ft^2-°F
Joule's law: ELECTRICAL

K

K: potassium (element)
Kelvin: relates to absolute temperature °C
kilo (k): prefix, equals 1000
kilocalorie: KILOGRAM-CALORIE
kilocycle: 1000 c
kilogram: metric unit of mass, the mass of a specific platinum/iridium standard kept at the International Bureau of Weights and Measures.
kilogram =

0.02205 cental	1 E + 06 mg
1 E + 05 cg	0.001 millier
1 E + 04 dg	0.1 myriagrams
257.21 dr (apoth/troy)	9.807 N
564.38 dr (avdp)	32.15 oz (apoth/troy)
9.80665 E + 05 dyn	35.274 oz (avdp)
1000 g	70.9316 pdl
0.0685 gee-lb	643 pwt
15,432.36 gr	0.01 quintal
10 hg	771.62 scruples (apoth)
0.09807 J/cm	0.0685 slug
9.807 J/m	0.001102 ton
2.679 lb (apoth/troy)	9.842 E − 04 ton (long)
2.205 lb (avdp)	0.001 ton (metric)

kilogram/cubic meter =

0.001 g/cm^3

0.06243 lb/ft^3

0.00835 lb/gal (liquid)

3.613 E − 05 lb/in.^{3}

3.405 E − 10 lb/mil-ft

1.6856 lb/yd^3

0.00194 slug/ft^3

kilogram/hectoliter =

0.01 g/cm^3

10 kg/m^3

0.62428 lb/ft^3

0.08345 lb/gal (liquid)

3.613 E − 04 lb/in.^{3}

16.855 lb/yd^3

kilogram/hour =

2.204 lb/h

0.0367 lb/min

kilogram/meter =

10 g/cm

0.672 lb/ft

0.056 lb/in.

9.807 N/m

1.774 ton/mi

1 ton (metric)/km

kilogram/meter/hour: metric unit of absolute (dynamic) viscosity

kilogram/meter/second: metric unit of absolute (dynamic) viscosity.

kilogram/meter/second: PASCAL-SECOND

kilogram/minute = 2.2046 lb/min

kilogram/square centimeter =

0.9678 atm

0.9807 bar

9.80663 E + 05 baryes

73.556 cm (mercury, 0°C, 32°F)

9.80665 E + 05 dyn/cm^2

32.808 ft (water, 4°C, 39.2°F)

1000 g/cm^2

28.959 in. (mercury, 0°C, 32 °F)

393.7 in. (water, 4°C, 39.2°F)

98.066 kPa

2048 lb/ft^2

14.223 lb/in.^{2}

0.73556 m (mercury, 0°C, 32°F)

735.56 mm (mercury, 0°C, 32°F)

1 E + 04 mm (water, 4°C, 39.2°F)

9.8066 E + 04 N/m^2

9.8066 E + 04 Pa

1.024 ton/ft^2

735.557 torr

kilogram/square meter =

9.678 E−05 atm	0.0098066 kPa
9.807 E−05 bar	0.2048 lb/ft^2
98.0662 baryes	0.001422 lb/in.^{2}
0.007356 cm (mercury, 0°C, 32°F)	0.07356 mm (mercury, 0°C, 32°F)
98.0662 dyn/cm^2	9.80662 N/m^2
0.003281 ft (water, 4°C, 39.2°F)	9.80662 Pa
0.1 g/cm^2	1.024 E−04 ton/ft^2
0.002896 in. (mercury, 0°C, 32°F)	7.117 E−07 ton/in.^{2}
0.03937 in. (water, 4°C, 39.2°F)	0.07356 torr
1 E−04 kg/cm^2	

kilogram/square meter-hour-°C = 0.2048 Btu/ft^2-h-°F

kilogram/square millimeter =

1 E+06 kg/m^2	9.807 MPa
2.048 E+05 lb/ft^2	0.7112 ton/in.^{2}
1422 lb/in.^{2}	

kilogram-calorie; calorie-kilogram; kilocalorie; French thermal unit: amount of heat required to raise the temperature of 1 kg of pure water 1°C.

kilogram-calorie =

3.9683 Btu	0.001580 hp-h (metric)
4.1293 E+04 cm^3-atm	4184 J
4.184 E+10 erg	426.7 kg-m
3086 ft-lb	4.184 kJ
9.9287 E+04 ft-pdl	0.0011624 kWh
1000 g-cal	41.29 l-atm
4.2665 E+07 g-cm	1.1624 Wh
0.001559 hp-h	4186 W-s

kilogram-calorie/hour =

3.968 Btu/h	0.001580 hp (metric)
0.001102 Btu/s	0.0011623 kW
0.001559 hp	1.1623 W

kilogram-calorie/hour-square meter-°C $= 0.2048$ Btu/h-ft^2-°F
kilogram-calorie/minute $=$

3.9683 Btu/min	0.09352 hp
0.06614 Btu/s	0.09482 hp (metric)
6.9737 E$+$08 erg/s	69.734 J/s
3086 ft-lb/min	0.0697 kW
51.467 ft-lb/s	69.734 W

kilogram-calorie-centimeter/hour-square meter-°C $=$

0.0806 Btu-in./h-ft^2-°F

kilogram-calorie-meter/hour-square meter-°C $= 8.06$ Btu-in./h-ft^2-°F
kilogram-calorie/second $=$

3.968 Btu/s	5.692 hp (metric)
4.184 E$+$10 erg/s	426.9 kg-m/s
3086 ft-lb/s	4.184 kW
5.611 hp	4184 W

kilogram-meter: metric unit for mechanical energy.
kilogram-meter $=$

0.00930 Btu	3.704 E$-$06 hp-h (metric)
9.80665 E$+$07 dyn-cm	9.80665 J
9.80665 E$+$07 erg	0.002342 kg-cal
0.003418 ft^3-atm	2.724 E$-$06 kWh
7.233 ft-lb	0.0967814 l-atm
232.715 ft-pdl	9.80665 N-m
2.344 g-cal	0.002724 Wh
1 E$+$05 g-cm	9.80665 W-s
3.653 E$-$06 hp-h	

kilogram-meter/gram $= 3280.84$ ft-lb/lb
kilogram-meter/hour $= 2.7241$ E$-$06 kW
kilogram-meter/minute $=$

0.00930 Btu/min	7.233 ft-lb/min
9.80665 E$+$07 erg/min	2.192 E$-$04 hp

kilogram-meter/minute — *continued*

2.22 E − 04 hp (metric)   1.63 E − 04 kW
0.002342 kg-cal/min   0.1634 W

kilogram-meter/second =

0.00930 Btu/s   0.01333 hp (metric)
9.80665 E + 07 erg/s   0.002342 kg-cal/s
7.233 ft-lb/s   0.009806 kW
0.01315 hp   9.80665 W

kilogram-second/square meter: metric unit of absolute (dynamic) viscosity.

kilogram-second/square meter =

9800 cP   9.8 E + 07 μP
98 dyn-s/cm^2 9800 mPa-s
98 g/cm/s   9.8 N-s/m^2
9.80 kg/m/s   98 P
0.205 lb (force)-s/ft^2 6.59 pdl-s/ft^2
0.00142 lb (force)-s/in.^{2} 0.205 slug/ft/s
6.59 lb (mass)/ft/s

kilojoule =

0.94845 Btu   0.2389 kg-cal
1000 J

kilojoule/kilogram = 0.43035 Btu/lb
kiloline =

1000 Mx
1 E − 05 Wb

kiloliter: CUBIC METER
kilometer =

6.689 E − 09 astronomical units   100 dam
1 E + 05 cm   3281 ft

kilometer — *continued*

10 hm	0.53996 mi (naut)
3.937 E + 04 in.	0.62137 mi (stat)
0.54 kn	3.937 E + 07 mils
0.18 league (naut)	1 E + 06 mm
0.207 league (stat)	0.1 myriameter
1.057 E − 13 light-years	198.84 rods
1000 m	1094 yd

kilometer/hour =

27.78 cm/s	0.5396 kn
3280.84 ft/h	16.67 m/min
54.68 ft/min	0.27778 m/s
0.9113 ft/s	0.6214 mi/h

kilometer/hour/second =

27.78 cm/s^2	0.2778 m/s^2
0.9113 ft/s^2	0.6214 mi/h/s

kilometer/minute =

1667 cm/s	60 km/h
1.96850 E + 05 ft/h	32.397 kn
3281 ft/min	37.28 mi/h
54.68 ft/s	0.62137 mi/min

kilometer/second = 37.28 mi/min

kilopascal =

0.009869 atm	0.2953 in. (mercury, 0°C, 32°F)
0.01 bar	4.0147 im. (water, 4°C, 39.2°F)
1 E + 04 baryes	0.010197 kg/cm^2
0.75006 cm (mercury, 0°C, 32°F)	101.97 kg/m^2
1 E + 04 dyn/cm^2	0.145 lb/in.^{2}
0.33456 ft (water, 4°C, 39.2°F)	0.0075 m (mercury, 0°C, 32°F)
10.197 g/cm^2	7.5006 mm (mercury, 0°C, 32°F)

kilopascal — *continued*

0.001 MPa	1000 Pa
1000 N/m²	7.5006 torr

kilovar: ELECTRICAL
kilovolt = 1000 V
kilovolt/cm =

1 E + 11 abvolts/cm	3.336 statvolts/cm
1 E + 11 μV/m	2540 V/in.
1 E + 08 mV/m	2.540 V/mil

kilovolt-ampere: ELECTRICAL
kilowatt: ELECTRICAL
kilowatt =

3414 Btu/h	1.341 hp
56.92 Btu/min	0.1019 hp (boiler)
0.9484 Btu/s	1.3596 hp (metric)
1 E + 10 erg/s	3.6 E + 06 J/h
1254.6 ft³-atm/h	1000 J/s
2.655 E + 06 ft-lb/h	860 kg-cal/h
4.4254 E + 04 ft-lb/min	14.34 kg-cal/min
737.6 ft-lb/s	0.239 kg-cal/s
1.42382 E + 06 ft-pdl/min	3.671 E + 05 kg-m/h
8.60375 E + 05 g-cal/h	102 kg-m/s
1.434 E + 04 g-cal/min	0.1 myriawatt
239 g-cal/s	1000 W
1.0197 E + 07 g-cm/s	

kilowatt/square foot = 56.904 Btu/min/ft²
kilowatt-hour: ELECTRICAL
kilowatt-hour =

3414 Btu	2.655 E + 06 ft-lb
3.6 E + 13 erg	8.543 E + 07 ft-pdl
1254.6 ft³-atm	8.60314 E + 05 g-cal

kilowatt-hour — *continued*

    3.671 E + 10 g-cm   860 kg-cal
    1.341 hp-h   3.671 E + 05 kg-m
    1.36 hp-h (metric)   3.5528 E + 04 l-atm
    3.6 E + 06 J   1000 Wh

kilowatt-hour/gram =

    1.548674 E + 06 Btu/lb   8.60375 E + 05 g-cal/g
    3.553 E + 07 cm^3-atm/g   608 hp-h/lb
    5.691 E + 05 ft^3-atm/lb   3.6 E + 06 J/g
    1.204 E + 09 ft-lb/lb

kilowatt-hour-centimeter/hour-square meter-°C =

    69.44 Btu-in./h-ft^2-°F

kilowatt-hour-inch/hour-square foot-°F =

    3414 Btu-in./h-ft^2-°F

kilowatt-hour-meter/hour-square meter-K =

    6944 Btu in./h-ft^2-°F

kinematics: study of distance and velocity of an object in motion, as in linear motion or rotation.

kinematic viscosities of various gases: GAS FLOW

kinematic viscosities of various liquids: FLUID FLOW

kinematic viscosity: VISCOSITY, KINEMATIC

kinetic energy: energy of a body in motion.

kinetic energy, linear motion: LINEAR MOTION

kinetic energy, rotation: ROTATION

kinetics: study of the action of a force causing or influencing the motion of a body.

kip: kilopound, 1000 pounds

Kirchhoff's law: ELECTRICAL

knot (length): MILE (NAUTICAL)
knot (speed) =

51.444 cm/s
6076 ft/h
101.27 ft/min
1.688 ft/s
1.852 km/h
0.030867 km/min
1852 m/h

30.87 m/min
0.5144 m/s
1 mi (naut)/h
0.01667 mi (naut)/min
1.151 mi (stat)/h
0.01918 mi (stat)/min
2025 yd/h

Kr: krypton (element)

L

La: lanthanum (element)
lambert: ILLUMINATION
lambert =

0.31831 cd/cm^2
295.72 cd/ft^2
2.054 cd/in.^{2}
3183 cd/m^2
929 ft-cd
929 ft L
1 lm/cm^2

929 lm/ft^2
1 E + 04 lm/m^2
1 E + 04 lx
1000 mL
1000 mph
1 ph

laminar flow: fluid (or gas) flowing in layers, as in nonturbulent flow, where the Reynold's number is less than 2000 by calculation.
latent heat: HEAT TRANSFER
law of cosines: TRIGONOMETRY
law of sines: TRIGONOMETRY
laws of thermodynamics: THERMODYNAMICS
league (nautical) =

3038 fathoms
18,228 ft
5.556 km

1.1508 leagues (stat)
3 mi (naut)
3.4523 mi (stat)

league (statute) =

2640 fathoms	0.869 league (naut)
15,840 ft	2.6069 mi (naut)
4.83 km	3 mi (stat)

LEL: lower explosive limit

Lenz's law: ELECTROMAGNETISM

leverage: the use of a lever arm and applied force for mechanical advantage, where

d = distance, ft
F = force, lb
W = weight, lb

force required,

$$F = Wd_1/d_2$$

lever arm: the perpendicular distance from a force to a point of rotation.

LHV: lower heating value

Li: lithium (element)

lighting terms: ILLUMINATION

light speed: VELOCITY OF LIGHT

light-year: the distance that light travels through space in a period of 1 yr.

light-year =

63,279 astronomical units	5.9 E + 12 mi (stat)
9.46 E + 12 km	

line: ELECTROMAGNETISM

line/square centimeter: GAUSS

line/square inch =

0.155 G	1 E − 08 Wb/in.^{2}
1.55 E − 09 Wb/cm^2	1.55 E − 05 Wb/m^2

line equivalent: MAXWELL
linear coefficient of expansion: HEAT TRANSFER
linear expansion, thermal: HEAT TRANSFER
LINEAR MOTION, rectilinear motion: where

a = acceleration, ft/sec^2
E = potential energy, ft-lb
F = force, lb
g = gravitational acceleration, 32.174 ft/sec^2
h = height, ft
hp = horsepower
I = impulse, lb-sec
K = kinetic energy, ft-lb
m = mass, lb-sec^2/ft, slugs
M = momentum, lb-sec
P = power, ft-lb/sec
s = displacement, ft
t = time, sec
t_1 = initial time, sec
t_2 = tinal time, sec
V = velocity, ft/sec
V_1 = initial velocity, ft/sec
V_2 = final velocity, ft/sec
w = weight, lb
W = work, ft-lb

**acceleration:**

$$a = V/t$$
$$= 0.5V^2/s$$
$$= (V_2 - V_1)/(t_2 - t_1)$$
$$= 0.5(V_2^2 - V_1^2)/s$$

displacement:

$$s = Vt$$
$$= 0.5V^2/a$$
$$= V_1 t + 0.5at^2$$
$$= 0.5(V_2^2 - V_1^2)/a$$

LINEAR MOTION — *continued*

(8) force:

$$F = ma$$
$$= m(V_2 - V_1)/t$$
$$= 0.5m(V_2^2 - V_1^2)/s$$
$$= 0.311wa$$
$$= 0.0311w(V_2 - V_1)/t$$
$$= 0.0155w(V_2^2 - V_1^2)/s$$
$$= W/s$$

(8) horsepower:

$$\text{hp} = FV/550$$
$$= P/550$$

impulse: force acted on a body over a time interval to cause a change in momentum of the body.

$$I = Ft$$
$$= F(t_2 - t_1)$$

kinetic energy: energy of a body due to its velocity,

$$K = 0.0155V^2w$$
$$= 0.5V^2m$$

mass:

$$m = w/g = 0.03108w$$

momentum:

$$M = mV$$
$$= 0.0311wV$$

Newton's laws of motion:

1. a body will remain at rest (or in uniform motion) until a force is applied to change the condition. The resultant of all forces acting on a body is zero.
2. a force acting on a body produces acceleration of the body parallel and directly proportional to the force, as $F = ma$.
3. every force acting on a body has a corresponding equal and opposing force.

LINEAR MOTION — *continued*

potential energy:

$$E = wh$$

(🔀) **power:**

$$P = W/t$$
$$= Fs/t$$
$$= FV$$

time:

$$t = s/V$$

(🔀) **velocity, linear:**

$$V = at = s/t$$
$$V_2 = V_1 + at$$
$$V_2 = (V_1^2 + 2as)^{1/2}$$

work: force moving through a distance; product of force and distance.

$$W = Fs$$
$$= Pt$$

---

link (engineer), link (Ramden), engineer's measure: FOOT
link (Gunter) =

0.01 chain (Gunter)	1 link (surveyor)
0.0066 chain (Ramden)	0.2012 m
20.1168 cm	1.25 E−04 mi (stat)
0.66 ft	0.04 rod
7.92 in.	

link (Ramden); link (engineer); engineer's measure: FOOT
link (surveyor): LINK (GUNTER)
liter: metric unit of volume equivalent to volume of 1 kg of water at its maximum density.

liter =

0.008387 bbl (liquid)	0.01 hl
0.00629 bbl (petroleum)	0.004193 hogshead
0.02838 bushel (level)	61.024 in.^{3}
100 cl	0.001 kl
1000 cm^3	0.001 m^3
0.1 dal	16,321 minims
1 E − 06 dam^3	1 E + 06 μl
1 E − 04 dekastere	1000 ml
10 dl	33.8 oz (liquid)
1 dm^3	0.1135 peck
270.512 dr (liquid)	1.8162 pt (dry)
0.035315 ft^3	2.1134 pt (liquid)
0.227 gal (dry)	0.90808 qt (dry)
0.220 gal (Imperial)	1.0567 qt (liquid)
0.2642 gal (liquid)	0.001 stere
8.454 gills	0.001308 yd^3

liter/minute =

0.035315 ft^3/min	0.004403 gal/s
5.886 E − 04 ft^3/s	0.06 m^3/h
0.26417 gal/min	1.667 E − 05 m^3/s

liter/second =

2.1189 ft^3/min	3.6 m^3/h
0.035315 ft^3/s	0.07848 yd^3/min
15.85 gal (liquid)/min	

liter-atmosphere =

0.0961 Btu	3.827 E − 05 hp-h (metric)
1.0132 E + 09 erg	101.328 J
0.035315 ft^3-atm	0.02422 kg-cal
74.74 ft-lb	10.333 kg-m
2404.55 ft-pdl	2.815 E − 05 kWh
24.217 g-cal	101.33 W-s
3.77 E − 05 hp-h	

liter-atmosphere/hour $= 0.0282$ W
ln: $\log_e$, natural logarithm, logarithm to base e
locked rotor amps: ELECTRICAL
log: $\log_{10}$; common logarithm; logarithm to base 10

LOGARITHMS:

basic function: The power (x) to which a specific base number (b) must be raised to equal a given number (N). As an example, x is the logarithm of number N to the base b, as

$$b^x = N, \quad \text{then } x = \log_b N$$

where $2^3 = 8$; 3 is the logarithm of 8 to the base 2.

common logarithms: where the number "10" is the base and noted as "log" or "$\log_{10}$." Tables of common logarithms of numbers can be found in various handbooks.

hyperbolic logarithms: natural logarithms   ∗

Napierian logarithms: natural logarithms   ∗

natural logarithms: hyperbolic logarithms, Napierian logarithms, where the letter e is the base ($e = 2.71828$) noted as "ln" or "$\log_e$." Tables of natural logarithms of numbers can be found in various handbooks. To convert from $\log_{10}$, $\ln = 2.30259 \log_{10}$.

usage of logarithms: logarithms of numbers consist of two parts, the characteristic and mantissa, separated by a decimal. As an example, using common logarithms,

    1. the characteristic identifies the number of digits the number has to the left of the decimal point *minus one*, as in the number 62, the characteristic $= 2 - 1 = 1$,

    2. the mantissa, a value obtained from logarithm tables, as a decimal value. For example, number 62 (620): mantissa $= .79239$. Then the complete logarithm for the number 62 is the combination of the mantissa and the characteristic, as 1.79239, or $10^{1.79239} = 62$.

    division of numbers, numbers can be divided by subtracting logarithms, as to divide 623 by 53,

$$\log 623 = 2.79449$$
$$\log 53 = 1.72428$$
$$\text{subtracting} = 1.07021$$

LOGARITHMS — *continued*

which is the logarithm of the unknown quotient.
The mantissa in 1.07021 is .07021.
The characteristic is 1.
Then in a reverse process, locating the number from the log tables that corresponds to this mantissa:

$$.07021 \text{ is } 1.1754$$

The quotient $= 1.1755$ with 1 place to the right of decimal,
$$= 11.754$$

multiplication of numbers, numbers can be multiplied by adding logarithms, as to multiply 623 by 53,

$$\log 623 = 2.79449$$
$$\log 53 = 1.72428$$
$$\text{adding} = 4.51877$$

which is the logarithm of the unknown product.
The mantissa in $4.51877 = .51877$.
The characteristic $= 4$.
Then in the reverse process, locating the number from the log tables that corresponds to this mantissa:

$$.51877 = 3.3019$$

The product $= 3.019$ with 4 places to right of the decimal,
$$= 33019.$$

powers of numbers: number can be raised to a power by multiplying the log of the number by the power, as $(25)^3$:

$$\log 25 = 1.39794$$
$$3(1.39754) = 4.19382$$

which is the logarithm of the answer.
The mantissa in 4.19382 is .19382.
The characteristic is 4.
Then in the reverse process, locating the number from the log tables that corresponds to this mantissa:

$$.19382 \text{ is } 1.5625$$

LOGARITHMS — *continued*

> The answer $= 1.5625$ with 4 places to the right of the decimal,
> $= 15625$.

---

logarithms to base e (ln), natural or Napierian logarithms: same procedure as logarithms to base 10 using natural or Napierian logarithm tables.

lohm: unit of resistance to flow of a fluid equal to passing 100 gal/min of 80°F water with a pressure drop of 25 psig.

lower heating value (LHV): refers to the heat value of a material when the water vapors formed by combustion remain in the products of combustion.

LRA: locked rotor amps.

Lu: lutetium (element)

lumen: ILLUMINATION

lumen $= 0.079577$ candlepower (spherical)

lumen/square centimeter: LAMBERT

lumen/square foot: FOOT-CANDLE; FOOT-LAMBERT

lumen/square meter: LUX

luminous flux: ILLUMINATION

lux: ILLUMINATION

lux $=$

0.0929 ft-cd	0.1 milliphot (mph)
0.0929 lm/ft^2	1 E$-$04 ph
1 lm/m^2	

M

mach number: ratio of velocity of an object to the speed of sound, where

> C = velocity of sound, 1100 ft/s
> M = Mach number
> V = velocity of the object, ft/s

> Mach number, $M = V/C$

MAGNETISM: pertaining to permanent magnets where metals such as hardened steel are magnetized and retain the magnetism, requiring no electricity for operation. Magnetic lines exit from the north pole and reenter the south pole of the magnet. In magnets, like poles repel and opposite poles attract, where

> A = area of pole face, cm^2
> B = flux density, gauss
> D = distance, cm
> f = magnetic flux, maxwell
> F = force, dyn
> H = field intensity, oersted
> p = pole strength, unit poles

Coulomb's law: the force between two magnetic poles is directly proportional to the product of the strength of the poles and inversely proportional to the square of the distance between the poles, as

$$F = p_1 p_2 / D^2$$

MAGNETISM — *continued*

field intensity: force per unit pole,

$$H = F/p$$

flux, magnetic: total number of magnetic lines in a magnetic field, as

$$f = 12.57p$$

flux density: the number of magnetic lines per square centimeter, expressed as gauss in the metric system,

$$B = f/A$$

force: product of pole strength and field intensity,

$$F = pH$$

pole strength; strength of a pole: equivalent to number of unit poles; a distance of 1 cm from a like pole repelling it with a force of 1 dyn.

---

magnetism, electro: ELECTROMAGNETISM

magnetomotive force: ELECTROMAGNETISM

manometer: tubular instrument formed to include a U section and containing a specific liquid (usually mercury or water); used to measure the difference between atmospheric pressure and pressure of a gaseous mixture flowing in a duct. Measurement is indicated on a vertical section of the manometer in inches of fluid.

manometer, inclined: basically, the same as the vertical manometer except readings are made on an inclined portion of the instrument. It is inclined to elongate the fluid reading for more accuracy than with the vertical reading.

marsh gas: methane

mass: quantity of inertia in a body, wherein inertia is proportional to the mass of the body,

English system, where

g = gravitational acceleration, 32.174 ft/s^2
m = mass, slugs
W = weight, lb

mass, $m = W/g = 0.03108W$
Metric system, mass units are grams or kilograms.

maxwell: metric unit for lines of induction or magnetic flux.
maxwell =

1 G-cm^2	3.336 E − 11 statweber
0.001 kiloline	1 E − 08 V-s
1 line	1 E − 08 Wb
1 E − 06 megaline	

maxwell/square centimeter: metric unit for intensity of magnetic induction.
maxwell/square centimeter: GAUSS
maxwell/square inch =

0.155 G	1 E − 08 Wb/in.^{2}
0.155 Mx/cm^2	

MCF natural gas = 1 E + 06 Btu, approx.
mechanical equivalent of heat: 1 Btu = 778.26 ft-lb of mechanical energy.
meg: MEGA
mega: prefix, equals 1 E + 06
megabar =

9.869 E + 05 atm
1 E + 06 bar

megacycle = 1 E + 06 c

megaline = 1 E + 06 Mx
megameter = 1 E + 06 m
megamho: MEGMHO
megapascal =

9.86923 atm	10.197 kg/cm^2
10 bar	0.1020 kg/mm^2
750 cm (mercury, 0°C, 32°F)	1000 kPa
1 E + 07 dyn/cm^2	145 lb/in.^{2}
295.3 in. (mercury, 0°C, 32°F)	1 E + 06 N/m^2
4014.7 in. (water, 4°C, 39.2°F)	1 E + 06 Pa

megawatt = 1E + 06 W
megmho, megamho =

0.001 abmho	1 E + 12 μmhos
1 E + 06 mhos	8.98758 E + 17 statmhos

megmho/centimeter; megmho/centimeter cube: metric unit for electrical
  conductivity per square centimeter of cross-sectional area of a
  material, in megmhos per centimeter length.
megmho/centimeter = 1 E + 06 mhos/cm
megmho/centimeter cube: MEGMHO/CENTIMETER
megmho/inch; megmho/inch cube: English unit of electrical conductivity
  per square inch of cross-sectional area of a material, in megmhos
  per inch of length.
megmho/inch = 1 E + 06 mhos/in.
megmho/inch cube: MEGMHO/INCH
megohm =

1 E + 15 abohms	1 E + 06 ohms
1 E + 12 microhms	1.113 E − 06 statohm

megohm-centimeter; megohm-centimeter cube: metric unit for electrical
  resistivity per square centimeter of cross-sectional area of a
  material, in megohms per centimeter length.
megohm-centimeter = 1 E + 06 ohm-cm

megohm-centimeter cube: MEGOHM-CENTIMETER

megohm-inch; megohm-inch cube: English units for electrical resistivity per square inch of cross-sectional area of a material, in megohms per inch length.

megohm-inch = 1 E + 06 ohm-inch

megohm-inch cube: MEGOHM-INCH

melting point of various materials (°F):

aluminum: 1220	copper: 1983
aluminum oxide: 3670	dolomite: 130
aluminum refractory: 3390	fire brick: 3000
antimony: 1166	galena: 2050
arsenic: 1500	gallium: 86
asphalt: 350	germanium: 1756
babbitt (lead type): 460	german silver: 1850
barium: 1500	glass: 2200
beeswax: 144	gold: 1940
beryllium: 2340	graphite: 6332
bismuth: 520	gypsum: 2480
borax: 1040	ice: 32
brass, red: 1952	iodine: 235
yellow: 1700	iron, cast: 2300
bronze, alum: 1850	malleable: 2850
phosphor: 1830	pure: 2800
zinc 10%: 1910	wrought: 2800
cadmium: 610	kaolin: 3200
calcium: 1560	lead: 621
calcium chloride: 1425	lead oxide: 1749
camphor: 350	lithium: 367
carbon: > 6300	magnesium: 1204
carnauba wax: 185	magnesium chloride: 1306
chrome ore: 3956	magnesium oxide: 5070
chromium: 2822	magnetite: 367
clay: 3160	manganese: 2300
cobalt: 2714	mercury: − 38
constantan: 2340	mercury chloride: 530

melting point of various materials (°F) — *continued*

molybdenum: 4750
monel, copper 30%: 2420
nichrome: 2552
nickel: 2650
palladium: 2830
paraffin: 133
petroleum jelly: 120
phenol: 106
phosphorus, red: 1095
           white: 111
pitch: 300
platinum: 3200
plutonium: 1170
porcelain: 2600
potassium: 145
radium: 1740
rhodium: 3600
rubber: 248
salt, rock: 1500
selenium: 428
shellac: 175
silica: 3182
silicon: 2590
silicon carbide: 4080
silver: 1760
sodium: 207
sodium carbonate: 1566

sodium chloride: 1472
sodium nitrate: 597
sodium sulfate: 1620
solder 50/50: 415
stearic acid: 157
steel, carbon: 2700
   silicon: 2690
   stainless: 2600
   structural: 2550
strontium: 1400
sugar: 320
sulfur: 235
tallow: 90
tantalum: 5160
tellurium: 846
thallium: 575
thorium: 3360
tin: 450
titanium: 3270
tungsten: 6100
uranium: 2070
vanadium: 3130
Wood's metal: 158
zinc: 788
zinc chloride: 690
zinc oxide: > 3240
zirconium: 3370

mesh: wire mesh; term used to denote size of screen openings, designating number of openings per inch, vertically and horizontally. The wire of the mesh displaces approximately half of the opening, as in a 100-mesh screen, the open area is $\frac{1}{2} \times 1/100$ in. and equals 0.5×0.01 in. $= 0.005$ in., or a 0.005×0.005 opening. One inch equals 25,400 microns; $0.005 \times 25,400 = 127$ microns, namely, a

mesh — *continued*

127 × 127 microns opening (actual manufacturer's rating is 149 microns for a 100-mesh screen, the difference being the actual wire size used and the $\frac{1}{2}$ estimate).

mesh sizes (openings in microns, μm):

20:	840	70:	210	200:	62	400:	30
30:	595	80:	177	270:	53	625:	20
40:	420	100:	149	300:	45	1250:	10
50:	297	140:	105	325:	44	2500:	5

meter: metric unit of length equal to the length of a specific bar of platinum-iridium at standard temperature and barometric pressure held at the International Bureau of Weights and Measures.

meter =

1 E + 10 Å	4.971 links (Gunter)
0.004557 cable lengths	3.2808 links (Ramden)
0.04971 chain (Gunter)	5.4 E − 04 mi (naut)
0.03281 chain (Ramden)	6.214 E − 04 mi (stat)
100 cm	39,370 mils
0.1 dam	1 E − 06 Mm
10 dm	1 E + 06 μm
0.5468 fathom	1000 mm
3.2808 ft	1 E + 09 nm
0.005 furlong	1 E + 12 pm
1 E − 09 gigameter	0.19884 rod
0.01 hm	0.00911 skein
39.37 in.	1.0936 yd
0.001 km	

meter (mercury, 0°C, 32°F) =

1.31579 atm	44.602 ft (water, 4°C, 39.2°F)
1.333 bar	1359.5 g/cm^2
1.333 E + 06 dyn/cm^2	39.37 in. (mercury, 0°C, 32°F)

meter (mercury, 0°C, 32°F) — *continued*

535.22 in. (water, 4°C, 39.2°F)   1000 mm (mercury, 0°C, 32°F)
1.3595 kg/cm^2 1.333 E + 05 Pa
133.322 kPa   1000 torr
19.3368 lb/in.^{2}

meter/hour =

3.2808 ft/h   5.4 E − 04 kn
0.05468 ft/min   6.2137 E − 04 mi/h

meter/minute =

1.667 cm/s   0.03239 kn
3.2808 ft/min   0.03728 mi/h
0.05468 ft/s   6.214 E − 04 mi/min
0.06 km/h

meter/second =

196.8 ft/min   1.9438 kn
3.2808 ft/s   2.237 mi/h
3.6 km/h   0.03728 mi/min
0.06 km/min

meter/second/second =

3.2808 ft/s^2 2.2374 mi/h/s
3.6 km/h/s

meter-kilogram: KILOGRAM-METER
meter-newton: JOULE
methane: marsh gas
methanol: methyl alcohol, wood alcohol
methyl alcohol: methanol, wood alcohol
Mg: magnesium (element)
mho; moh: unit of electrical conductance

mho =

1 E−09 abmho	1/ohm
1 E−06 megmho	1 Siemen unit
1 E+06 micromos	8.98758 E+11 statmhos

mho/centimeter; mho/centimeter cube: metric unit for electrical conductivity per square centimeter of cross-sectional area of a material, in mhos per centimeter of length.

mho/centimeter =

1 E−09 abmho/cm	1.66 E−07 mho-ft/mil
1 E−06 megmho/cm	1 E−04 mho-m/mm²
2.54 E−06 megmho/in.	1 E+06 micromhos/cm
2.54 mhos/in.	2.54 E+06 micromhos/in.
100 mhos/m	8.98758 E+11 statmhos/cm

mho/centimeter cube: MHO/CENTIMETER
mho/circular mil/foot: MHO-FOOT/MIL
mho/inch; mho/inch cube: English unit for electrical conductivity per square inch of cross-sectional area of a material, in mhos per inch of length.

mho/inch; mho/inch cube =

3.937 E−10 abmho/cm	0.3937 mho/cm
3.937 E−07 megmho/cm	39.37 mhos/m
1 E−06 megmho/in.	3.937 E+05 micromhos/cm
6.545 E−08 mho-ft/mil	1 E+06 micromhos/in.
3.397 E−05 mho-m/mm²	3.495 E+11 statmhos/cm

mho/inch cube: MHO/INCH
mho/meter; mho/meter cube: metric unit for electrical conductivity per square meter of cross-sectional area of a material, in mhos per meter of length.

mho/meter = 0.01 mho/cm
mho/meter cube: MHO/METER
mho/meter/gram: metric unit of electrical conductivity for each gram weight per meter of a material, in mhos per meter of length.

mho/meter/gram = 5710 mhos/mi/lb

mho/mile/pound: English unit for electrical conductivity for each pound weight per mile of a material, in mhos per mile.

mho/mile/pound = 1.75 E − 04 mho/m/g

mho-foot/circular mil: MHO-FOOT/MIL

mho-foot/mil; mho/circular mil/foot; mho-foot/circular mil; mho-mil-foot: English unit for electrical conductivity per circular mil cross-sectional area of a material, in mhos per foot of length.

mho-foot/mil; mho/circular mil/foot =

0.006015 abmho/cm	6.015 E + 08 mhos/m
6.015 megmhos/cm	601.5 mhos-m/mm²
15.28 megmhos/in.	6.0153 E + 12 micromhos/cm
6.015 E + 06 mhos/cm	1.528 E + 13 micromhos/in.
1.528 E + 07 mhos/in.	5.406 E + 18 statmhos/cm

mho-meter/square millimeter: metric unit for electrical conductivity per square millimeter cross-sectional area of a material, in mhos per meter of length.

mho-meter/square millimeter =

1 E + 05 ambmho/cm	1 E + 06 mhos/m
0.01 megmho/cm	0.00166 mho-ft/mil
0.0254 megmho/in.	1 E + 10 micromhos/cm
1 E + 04 mhos/cm	2.54 E + 10 micromhos/in.
2.54 E + 04 mhos/in.	8.98758 E + 15 statmhos/cm

mho-mil-foot: MHO-FOOT/MIL

micro (μ): prefix, equals 1 E − 06

microampere =

1 E − 06 A	0.001 mA

microbar: DYNE/SQUARE CENTIMETER

microcoulomb = 1 E − 06 C

microfarad =

1 E − 15 abfarad	8.98758 E + 05 statfarads
1 E − 06 F	

microgram =

   1 E−06 g   0.001 mg

microhenry =

   1 E−06 H   1.113 E−18 stathenry
   0.001 mH

microhm =

   1000 abohms   1 E−06 ohm
   1 E−12 megohm   1.113 E−18 statohm

microhm-centimeter; microhm-centimeter cube: metric unit for electrical resistivity per square centimeter of cross-sectional area of a material, in microhoms per centimeter length.

microhm-centimeter = 1 E−06 ohm-cm

microhm-centimeter cube: MICROHM-CENTIMETER

microhm-inch; microhm-inch cube: English unit for electrical resistivity per square inch of cross-sectional area of a material, in microhms per inch length.

microhm-inch = 1 E−06 ohm-in.

microhm-inch cube: MICROHM-INCH

microinch =

   1 E−06 in.
   2.54 E−08 m

microliter = 1 E−06 l

micrometer: MICRON

micromho =

   1 E−15 abmho   1 E−06 mho
   1 E−12 megmho   8.98758 E+05 statmhos

micromho/centimeter; micromho/centimeter cube: metric unit for electrical conductivity per square centimeter of cross-sectional area of a material, in micromhos per centimeter of length.

micromho/centimeter = 1 E−06 mho/cm
micromho/centimeter cube: MICROMHO/CENTIMETER
micromho/inch; micromho/inch cube: English unit for electrical conductivity per square inch of cross-sectional area of a material, in micromhos per inch of length.
micromho/inch = 1 E−06 mho/in.
micromho/inch cube: MICROMHO/INCH
micromicro: prefix, equals 1 E−12
micromicrofarad =

> 1 E−12 F
> 1 E−06 μF

micromicron =

> 0.01 Å   1 E−12 m
> 1 E−10 cm   1 E−06 μm
> 3.937 E−11 in.

micron =

> 1 E+04 Å   1 E−06 m
> 1 E−04 cm   1 μm
> 3.2808 E−06 ft   0.001 mm
> 3.937 E−05 in.   1000 nm

micron (mercury, 0°C, 32°F) =

> 1.316 E−06 atm   0.0136 kg/m^2
> 1.333 E−06 bar   1.33 E−04 kPa
> 1 E−04 cm (mercury)   1.934 E−05 lb/in.^{2}
> 1.333 dyn/cm^2 0.00278 lb/ft^2
> 0.00136 g/cm^2 0.001 mm (mercury)
> 3.937 E−05 in. (mercury)   0.1333 N/m^2
> 5.353 E−04 in. (water, 4°C, 39.2°F)   0.1333 Pa
> 1.36 E−06 kg/cm^2

micropoise: metric unit of absolute (dynamic) viscosity.

micropoise =

1 E−04 cP
1 E−06 dyn-s/cm^2
1 E−06 g/cm/s
1 E−07 kg/m/s
1.02 E−08 kg-s/m^2
2.09 E−09 lb (force)-s/ft^2
1.45 E−11 lb (force)-s/in.^{2}

6.72 E−08 lb (mass)/ft/s
1 E−04 mPa-s
1 E−07 N-s/m^2
1 E−06 P
6.72 E−08 pdl-s/ft^2
2.09 E−09 slug/ft/s

microvolt =

100 abvolts
1 E−06 V

mil: one-thousandth of an inch; also used as an abbreviation for the circular mil.

mil =

0.00254 cm
8.333 E−05 ft
0.001 in.
2.54 E−08 km

2.54 E−05 m
0.0254 mm
2.778 E−05 yd

mile, mile (statute) =

1.609 E+05 cm
80 chains (Gunter)
52.8 chains (Ramden)
880 fathoms
5280 ft
8 furlongs
6.336 E+04 in.
1.609 km
0.8688 kn
0.2897 league (naut)
0.3333 league (stat)

1.70 E−13 light-year
8000 links (Gunter)
5280 links (Ramden)
1.6093 E+03 m
0.869 mi (naut)
1 mi (stat)
1.6093 E+06 mm
0.1609 myriameter
320 rods
1760 yd

mile (nautical): knot, established as 1 min (angular) of arc of the earth's surface.

mile (nautical) =

8.43905 cable lengths	1 kn
1.852 E+05 cm	0.3333 league (naut)
1012.686 fathoms	0.3836 league (stat)
6076.1155 ft	1852 m
9.206 furlongs	1.15078 mi (stat)
1.852 km	2026.7 yd

mile (nautical)/hour: KNOT (SPEED)
mile (statute): MILE
mile/hour =

44.70 cm/s	0.02682 km/min
5280 ft/h	0.8684 kn
88 ft/min	1609 m/h
1.467 ft/s	26.822 m/min
6.336 E+04 in./h	0.447 m/s
1056 in./min	0.01667 mi/min
1.609 km/h	

mile/hour/hour = $4.074 \text{ E}-04 \text{ ft/s}^2$
mile/hour/minute = 0.7451 cm/s^2
mile/hour/second =

44.704 cm/s²	1.6094 km/h/s
1.467 ft/s²	0.44704 m/s²

mile/minute =

2682 cm/s	1.6093 km/min
3.168 E+05 ft/h	0.026824 km/s
5280 ft/min	52.14 kn
88 ft/s	1609 m/min
6.336 E+04 in./min	26.824 m/s
1056 in./s	60 mi/h

mile/second =

> 1.60934 E + 05 cm/s
> 1609.344 m/s

mil-ft: CIRCULAR MIL-FT
milli: prefix, equals 1 E − 03
milliampere =

> 0.001 A
> 1000 μA

millibar =

> 9.869 E − 04 atm
> 0.001 bar
> 1000 baryes
> 0.075006 cm (mercury, 0°C, 32°F)
> 1000 dyn/cm^2
> 1.0197 g/cm^2
> 0.02953 in. (mercury, 0°C, 32°F)
> 0.4015 in. (water, 4°C, 39.2°F)
>
> 0.0010197 kg/cm^2
> 0.1 kPa
> 2.0885 lb/ft^2
> 0.0145 lb/in.^{2}
> 0.75006 mm (mercury, 0°C, 32°F)
> 100 N/m^2
> 100 Pa
> 0.75006 torr

millier: TON (METRIC)
milligram =

> 0.004871 carat
> 0.1 cg
> 2.572 E − 04 dr (apoth/troy)
> 5.6438 E − 04 dr (avdp)
> 0.001 g
> 0.0154 gr
> 1 E − 06 kg
>
> 2.679 E − 06 lb (apoth/troy)
> 2.2046 E − 06 lb (avdp)
> 1000 μg
> 3.215 E − 05 oz (apoth/troy)
> 3.5274 E − 05 oz (avdp)
> 6.430 E − 04 pwt
> 7.7162 E − 04 scruple (apoth)

milligram/centimeter =

> 0.980665 dyn/cm
> 5.6 E − 06 lb/in.

milligram/inch =

0.386089 dyn/cm	3.937 E−04 g/cm
0.980665 dyn/in.	0.001 g/in.

milligram/liter: PART/MILLION
milligram/millimeter − 9.80665 dyn/cm
millihenry =

1 E+06 abhenrys	1000 μH
0.001 H	1.111 E−15 stathenry

millilambert =

3.183 E−04 cd/cm²	0.929 lm/ft²
0.00205 cd/in.²	10 lm/m²
3.183 cd/m²	10 lx
0.929 ft-L	1 milliphot (mph)
0.001 L	0.001 ph
0.001 lm/cm²	

milliliter: CUBIC CENTIMETER
milliliter/second: CUBIC CENTIMETER/SECOND
millimeter =

1 E+07 Å	0.001 m
0.1 cm	39.37 mils
1 E−04 dam	6.214 E−07 mi (stat)
0.01 dm	1000 μm
0.003281 ft	1 E+06 nm
0.03937 in.	0.001094 yd
1 E−06 km	

millimeter/gram = 1.602 ft³/lb
millimeter/second = 0.03937 in./s
millimeter (mercurcy, 0°C, 32°F): TORR
millimeter (water, 4°C, 39.2°F) = 9.678 E−05 atm
millimicro: prefix, equals 1 E−09 (same as nano)

millimicron: NANOMETER
millipascal-second: metric unit of absolute (dynamic) viscosity.
millipascal-second: CENTIPOISE
milliphot: MILLILAMBERT
millivolt =

    1 E + 05 abvolts   0.001 V
    3.336 E − 06 statvolt

mil-ohm-foot: OHM-MIL/FOOT
minim =

    0.0616 cm^3 0.0616 ml
    0.01667 dr (liquid)   61.61 mm^3
    1.63 E − 05 gal (liquid)   0.002083 oz (liquid)
    5.21 E − 04 gill   1.3 E − 04 pt (liquid)
    0.00376 in.^{3} 6.5 E − 05 qt (liquid)
    6.127 E − 05 l

minute =

    0.01667 h
    60 s

minute (angular) =

    0.01667°   60 s (angular)
    2.909 E − 04 rad

mks: meter-kilogram-second
MMBTU: million Btu; approximately 1000 ft^3 natural gas
Mn: manganese (element)
Mo: molybdenum (element)
modulus of elasticity: STRENGTH OF MATERIALS
modulus of rigidity: STRENGTH OF MATERIALS
modulus of rupture: STRENGTH OF MATERIALS
modulus of shear (G): STRENGTH OF MATERIALS
moh: MHO

mol; mole: GRAM MOLECULAR WEIGHT

molecular weight: sum of the weights of the individual atoms in a molecule.

molecular weight of various gases and liquids:

acetic acid: 60.05
acetone: 58.08
acetylene: 26.03
air: 28.9752
ammonia: 17.03
argon: 39.90
benzene: 78.05
butane: 58.08
carbon dioxide: 44.01
carbon monoxide: 28.01
carbon tetrachloride: 153.8
chlorine: 70.91
chloroform: 119.39
ethane: 30.07
ether: 74.125
ethyl alcohol: 46.07
ethylene: 28.03
ethylene glycol: 62.07
freon F12: 120.9
    F22: 92.09

gasoline: 86
helium: 4.0
heptane: 100.21
hydrogen: 2.016
hydrogen choride: 36.47
hydrogen fluoride: 20.01
hydrogen sulfide: 34.08
methane: 16.03
methyl alcohol: 32.04
naphtha: 106.16
nitrogen: 28.02
octane: 114.235
oxygen: 32.0
propane: 44.098
sulfur dioxide: 64.07
toluene: 92.14
turpentine: 136.24
water: 18.01
xylene: 106.16

molecule: combination of two or more atoms.

moment: STATICS

moment of inertia, angular: ROTATION

MOMENT OF INERTIA, plane area: where

A = area, in.^{2}
d = distance between axes, in.
I = moment of inertia, in.^{4}
I_P = polar moment of inertia, in.^{4}
k = radius of gyration, in.

MOMENT OF INERTIA, plane area — *continued*

moment of inertia for various plane area sections; refer to: ANGLE; CHANNEL; CIRCLE; ELLIPSE; HEXAGON; OCTAGON; PARABOLA; RECTANGLE; SQUARE; TEE; TRAPEZOID; TRIANGLE; Z BAR.

moment of inertia relating to an axis in the same plane: it is the summation of the products of each particle area of the surface multiplied by the square of the distance of the particle area to the referenced axis. Also a function of the radius of gyration.

$$I = k^2 A$$

polar moment of inertia: relating to an axis through the center of gravity and perpendicular to the plane of the area; it is the summation of each particle area of the surface multiplied by the square of the distance to the referenced axis. Also the sum of moments of inertia about the x and y axes of the plane area.

$$I_P = I_x + I_y$$

transfer moment of inertia: transfer to a parallel axis I_2 of a distance d from referenced axis I_1,

$$I_2 = I_1 + d^2 A$$

moment of inertia, solid: where

d = distance between axes, ft
I = moment of inertia, slug-ft^2
m = mass, slugs
W = weight, lb

moment of inertia for various solids, refer to: CONE; CUBE; CYLINDER; DISK; ELLIPSOID; PYRAMID; RECTANGULAR PRISM; RING; ROD; SPHERE.

moment of inertia in units of mass (slug-ft^2): relating to a referenced axis; it is the summation of the products of each particle of mass multiplied by the square of the distance of each particle to the axis. For units of weight (lb-ft^2), multiply the moment of inertia units of mass by 32.174.

transfer moment of inertia: transfer to a parallel axis I_2 a distance d from the original referenced axis I_1,

$$I_2 = I_1 + d^2 m$$
$$= I_1 + 0.03108 d^2 W$$

momentum: LINEAR MOTION
momentum, angular: ROTATION
month =

30.4167 d	43,800 min
730 h	

Moody diagram: chart for determining friction factor of air, fluids, gases, or steam flowing through pipes based on specific conditions of flow, pipe size, pipe roughness, and Reynolds number. Charts can be found in various handbooks.

motors, electric: ELECTRICAL
mPa: millipascal
MPa: megapascal
MV: megavolts
MW: molecular weight
myria: prefix, equals 1 E + 04
myriagram =

1 E + 04 g	22.046 lb (avdp)
10 kg	352.74 oz (avdp)

myriameter =

10 km	6.2137 mi (statute)
5.3995 mi (nautical)	

myriawatt = 10 kW

N

N: nitrogen (element)
Na: sodium (element)
nano (n): prefix, equals 1 E−09 (same as millimicro)
nanometer =

10 Å
1 E−07 cm
3.937 E−08 in.
1 E−09 m

0.001 μm
1 E−06 mm
1 millimicrometer

Napierian logarithm: LOGARITHMS
natural gas: by composition,

83.4% methane (CH₄)
15.8% ethane (C₂H₆)

0.8% nitrogen (N₂)

natural logarithm: LOGARITHMS
nautical mile: MILE, NAUTICAL
Nb: niobium (element)
NCMH: normal cubic meters per hour
Nd: neodymium (element)
Ne: neon (element)
NEMA: National Electrical Manufacturer's Association
Neper = 8.686 decibels
neutron: neutral part of an atom nucleus consisting of an electron and
 proton.
neutron mass =

1.67 E−24 g

1.67 E−27 kg

newton: metric unit of force that accelerates 1 kg of mass at 1 m/s^2.
newton =

1 E+05 dyn	0.10197 kg
101.967 g	0.2248 lb (force)
0.01 J/cm	7.233 pdl
1 J/m	0.000102 ton (metric)

newton/meter =

0.0102 g/cm	0.06854 lb (force)/ft
0.10197 kg/m	

newton/square meter: PASCAL
newton-meter: metric unit for work
newton-meter: JOULE
newton-second/square meter: metric unit of absolute (dynamic) viscosity.
newton-second/square meter: PASCAL-SECOND
Newton's Laws of motion: LINEAR MOTION
Ni: nickel (element)
N$_{NU}$: NUSSELT NUMBER
noise levels for various sounds: SOUND
normal acceleration: ROTATION
normal cubic meters per hour (NCMH): AIR FLOW
nozzle flow: FLUID FLOW
Np: neptunium (element)
N$_R$: REYNOLDS NUMBER
nucleus: positively charged stationary part of an atom.
numbers of elements: ELEMENTS
Nusselt number (N$_{Nu}$): dimensionless number used in fluid flow and convection heating calculations, where

D = diameter of pipe, ft
h = heat transfer coefficient, Btu/hr-ft^2-°F
k = thermal conductivity, Btu-in./hr-ft^2-°F
N_{NU} = Nusselt number

Nusselt number,

$$N_{Nu} = hD/k$$

O

O: oxygen (element)

obtuse angle: an angle larger than 90° and less than 180°.

OCTAGON SECTION: where

a, h, x, y = dimensions, in.

A = cross-sectional area, in.^{2}

cg = center of gravity

I_{AA}, I_{BB} = moment of inertia, in.^{4}

k_{AA}, k_{BB} = radius of gyration, in.

Z_{AA}, Z_{BB} = section modulus, in.^{3}

area,

$$A = 4.828a^2 = 0.828h^2$$

center of gravity (cg),

$$x = 0.5h = 1.207a$$
$$y = 0.5h = 1.207a$$

distance,

$$a = 0.414h$$
$$h = 2.414a$$

moment of inertia,

$$I_{AA} = I_{BB} = 0.055h^4$$

radius of gyration,

$$k_{AA} = k_{BB} = 0.257h$$

OCTAGON SECTION — *continued*

section modulus,

$$Z_{AA} = Z_{BB} = 0.109h^3$$

---

oersted: metric unit for magnetizing force or magnetic field intensity in the electromagnetic system.

oersted =

0.0796 abampere-turn/cm	79.577 A-turns/m
0.796 A-turn/cm	1 Gb/cm
2.021 A-turns/in.	3 E + 10 statoersteds

ohm: ELECTRICAL

ohm =

1 E + 09 abohms	1 E + 06 microhms
1 E − 06 megohm	1.113 E − 12 statohm
1/mho	

ohm (centimeter cube): OHM-CENTIMETER
ohm (circular mil-foot): OHM-MIL/FOOT
ohm/centimeter cube: OHM-CENTIMETER
ohm/inch cube: OHM-INCH
ohm/meter cube: OHM-METER
ohm/mile pound: OHM-POUND/MILE
ohm/mil-foot: OHM-MIL/FOOT
ohm-centimeter; ohm/centimeter cube; ohm-centimeter cube; ohm (centimeter cube): metric unit of electrical resistivity in ohms per centimeter length for a material with cross-sectional area of 1 cm^2.

ohm-centimeter =

1 E + 09 abohm-cm	0.3937 ohm-in.
1 E − 06 megohm-cm	0.01 ohm-m
3.937 E − 07 megohm-in.	6.015 E + 06 ohm-mil/ft
1 E + 06 microhm-cm	1 E + 04 ohm-mm^2/m
3.937 E + 05 microhm-in.	1.113 E − 12 statohm-cm

ohm-centimeter cube: OHM-CENTIMETER

ohm-gram/meter; ohm (meter-gram): metric unit of electrical resistivity in ohms per meter length for a material weighing 1 g per meter of length.

ohm-gram/meter $= 1.75$ E $- 04$ ohm-lb/mi

ohm-inch; ohm/inch cube; ohm (in cube); ohm-inch cube: English unit of electrical resistivity in ohms per inch length for a material with cross-sectional area of 1 in.^{2}

ohm-inch $=$

2.54 E+09 abohm-cm	2.54 ohm-cm
2.54 E−06 megohm-cm	0.0254 ohm-m
1 E−06 megohm-in.	1.528 E+07 ohm-mil/ft
2.54 E+06 microhm-cm	2.54 E+04 ohm-mm^2/m
1 E+06 microhm-in.	2.83 E−12 statohm-cm

ohm (inch cube): OHM-INCH

ohm-inch cube: OHM-INCH

ohm-meter; ohm/meter cube; ohm (meter cube); ohm-meter cube: metric unit of electrical resistivity in ohms per meter length for a material with cross-sectional area of 1 m^2.

ohm-meter $= 100$ ohm-cm

ohm-meter cube: OHM-METER

ohm (meter cube): OHM-METER

ohm (meter-gram): OHM-GRAM/METER

ohm (meter-square millimeter): OHM-SQUARE MILLIMETER/METER

ohm-mil/foot; circular mil-ohm/foot; mil-ohm-foot; ohm (circular mil-foot); ohm (mil-foot); ohm/mil-foot; ohm-mil-foot: English unit of electrical resistivity in ohms per foot length for a material with cross-sectional area of 1 circular mil.

ohm-mil/foot $=$

166 abohm-cm	0.166 microhm-cm
1.662 E−13 megohm-cm	0.06545 microhm-in.
6.545 E−14 megohm-in.	1.66 E−07 ohm-cm

ohm-mil/foot — *continued*

0.01478 ohm-g/m	1.66 E − 09 ohm-m
6.545 E − 08 ohm-in.	0.00166 ohm-mm²/m
84.389 ohm-lb/mi	1.85 E − 19 statohm-cm

ohm-mil-foot: OHM-MIL/FOOT
ohm (mil-foot): OHM-MIL/FOOT
ohm (mile-pound): OHM-POUND/MILE
ohm-pound/mile; ohm/mile-pound; ohm (mile-pound): English unit of electrical resistivity in ohms per mile length for a material weighing 1 pound per mile length.
ohm-pound/mile = 5710 ohm-g/m
Ohm's law: ELECTRICAL
ohm-square millimeter/meter; ohm (meter-square millimeter): metric unit of electrical resistivity in ohms per meter length for a material with cross-sectional area of 1 mm².
ohm-square millimeter/meter =

1 E + 05 abohm-cm	1 E − 04 ohm-cm
1 E − 10 megohm cm	3.937 E − 05 ohm-in.
3.937 E − 11 megohm-in.	1 E − 06 ohm-m
100 microhm-cm	601.5 ohm-mil/ft
39.37 microhm-in.	1.113 E − 16 statohm-cm

orifice: FLUID FLOW
Os: osmium (element)
ounce =

138 carats (metric)	2.83495 E + 04 mg
7.292 dr (apoth/troy)	0.002835 myriagrams
16 dr (avdp)	0.9115 oz (apoth/troy)
28.35 g	28.35 ponds
437.5 gr	18.23 pwt
6.25 E − 04 hwt	21.875 scruples (apoth)
0.02835 kg	3.125 E − 05 ton
0.0625 lb	2.835 E − 05 ton (metric)
0.07595 lb (apoth/troy)	

ounce/centimeter $= 0.15875$ lb/in.^{2}
ounce/cubic foot $=$

0.0625 lb/ft^3	5.787 E-04 oz/in.^{3}
0.008355 lb/gal	3.125 E-05 ton/ft^3
3.6169 E-05 lb/in.^{3}	8.4375 ton/yd^3
1.6875 lb/yd^3	

ounce/cubic inch $=$

1730 kg/m^3	2916 lb/yd^3
108 lb/ft^3	1728 oz/ft^3
14.438 lb/gal	0.054 ton/ft^3
0.0625 lb/in.^{3}	1.458 ton/yd^3

ounce/gallon $= 7.4892$ kg/m^3
ounce/square foot $=$

8.884 E-04 in. (mercury, 0°C, 32°F)	4.34 E-04 lb/in.^{2}
0.012 in. (water, 4°C, 39.2°F)	0.00694 oz/in.^{2}
0.0625 lb/ft^2	

ounce/square inch $=$

0.00425 atm	9 lb/ft^2
4309.22 dyn/cm^2	0.0625 lb/in.^{2}
0.106 ft (mercury, 0°C, 32°F)	3.23 mm (mercury, 0°C, 32°F)
0.1443 ft (water, 4°C, 39.2°F)	44.02 mm (water, 4°C, 39.2°F)
4.394 g/cm^2	144 oz/ft^2
0.12725 in. (mercury, 0°C, 39.2°F)	431 Pa
1.73 in. (water, 4°C, 39.2°F)	

ounce (apoth/troy) $=$

3.11 dag	0.06857 lb (avdp)
8 dr (apoth/troy)	3.11035 E$+04$ mg
31.103 g	1.0971 oz (avdp)
480 gr	20 pwt
0.0311 kg	24 scruples (apoth)
0.0833 lb (apoth/troy)	3.4286 E-05 ton

ounce (force) = 0.278014 N
ounce (liquid) =

8.392 E−04 bushel (level)	2.957 E−04 hl
2.957 cl	0.02957 l
29.574 cm^3	2.9574 E−05 m^3
0.125 cup	29.574 ml
8 dr (liq)	480 minims
1.8047 in.^{3}	1.0408 oz (liq, Brit.)
0.0010444 ft^3	0.0625 pt (liq)
0.006714 gal (dry)	0.03125 qt (liq)
0.007813 gal (liq)	2 tbsp
0.25 gill	6 tsp

ounce (liq, apoth) =

1.805 in.^{3}	0.0078 gal
8 dr	

ounce (liq, Brit.) =

1.734 in.^{3}
0.9608 oz (liq)

ounce (troy) =

155.5 carats	0.06857 lb (avdp)
31.103 g	0.0833 lb (troy)
480 gr	1.09714 oz (avdp)
0.0311 kg	20 pwt

ounce-inch: INCH-OUNCE
overall coefficient of heat transfer (U): HEAT TRANSFER
oz: ounce

P

P: phosphorus (element)

Pa: protactinium (element)

PARABOLA: cross-section of a cone taken through a plane below the cone vertex and parallel to the side of the cone, where

$a = FV = VC$

d = distance from point F to any point b

   = perpendicular distance from DD to point b

DD = directrix, parallel to EE

EE = latus rectum = $4FV$

  F = focus

FC = perpendicular to DD

FC = FE

  V = vertex

curve equation,

$$y^2 = 4ax \quad \text{or} \quad x^2 = 4ay$$

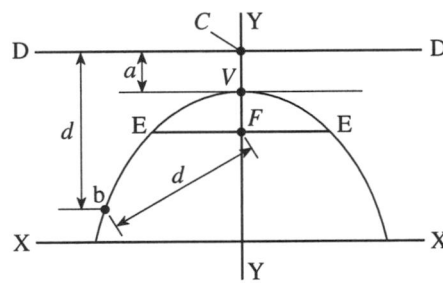

PARABOLA SECTION: where

h, L, x, y = dimensions, in.

A = cross-sectional area, in.^{2}

cg = center of gravity

$I_{AA}, I_{BB}, I_{CC}, I_{DD}$ = moment of inertia, in.^{4}

$k_{AA}, k_{BB}, k_{CC}, k_{DD}$ = radius of gyration, in.

s = arc length, in.

PARABOLA SECTION — *continued*

arc length,

$$s = 0.5(16h^2 + L^2)^{1/2} + 0.125(L^2/h) \ln [4h + (16h^2 + L^2)^{1/2}/L]$$
$$= 2(x^2 + 1.33h^2)^{1/2} \text{ approx.}$$

area,

$$A = 0.667Lh$$

center of gravity,

$$x = 0.5$$
$$y = 0.4h$$

moment of inertia,

$$I_{AA} = 0.0457h^3L$$
$$I_{BB} = 0.0333L^3h$$
$$I_{CC} = 0.2857h^3L$$
$$I_{DD} = 0.1524h^3L$$

radius of gyration,

$$k_{AA} = 0.2619h$$
$$k_{BB} = 0.2235L$$
$$k_{CC} = 0.6546h$$
$$k_{DD} = 0.4781h$$

PARALLELOGRAM: four-sided figure with opposite sides parallel and equal.

area,

$$A = ah = ab \sin \theta$$
$$c = (a^2 + b^2 - 2ab \cos \theta)^{1/2}$$
$$d = (a^2 + b^2 + 2ab \cos \theta)^{1/2}$$
$$h = b \sin \theta$$

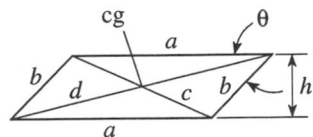

center of gravity (cg), at intersection of diagonals

part/million =

35.3147 g/ft^3

0.001 g/l

1 g/m^3

0.437 gr/ft^3

part/million — *continued*

0.0584 gr/gal	6.243 E−05 lb/ft³
0.07015 gr/gal (Imperial)	1 mg/l

pascal =

9.869 E−06 atm	0.10197 kg/m²
1 E−05 bar	0.001 kPa
10 baryes	0.02088 lb/ft²
7.500 E−04 cm (mercury, 0°C, 32°F)	1.45 E−04 lb/in.²
10 dyn/cm²	0.0075 mm (mercury, 0°C, 32°F)
3.345 E−04 ft (water, 4°C, 39.2°F)	0.10197 mm (water, 4°C, 39.2°F)
0.010197 g/cm²	1 E−06 MPa
2.953 E−04 in. (mercury, 0°C, 32°F)	1 N/m²
0.004014 in. (water, 4°C, 39.2°F)	0.00231 oz/in.²
1.0197 E−05 kg/cm²	0.0075 torr

Pascal's law: when pressure is applied to a fluid enclosed in a container, the pressure is transmitted to every portion of the fluid and walls of the container vessel.

pascal-second: metric unit of absolute (dynamic) viscosity.

pascal-second =

1000 cP	0.672 lb (mass)/ft/s
10 dyn-s/cm²	1 E+07 μP
10 g/cm/s	1000 mPa-s
1 kg/m/s	1 N-s/m²
0.102 kg-s/m²	10 P
0.0209 lb (force)-s/ft²	0.672 pdl-s/ft²
1.45 E−04 lb (force)-s/in.²	0.0209 slug/ft/s

Pb: lead (element)
Pd: palladium (element)
peck =

0.07619 bbl (dry)	0.881 dal
0.25 bushel (level)	0.3111 ft³
8810 cm³	2 gal (dry)

peck — *continued*

2.3273 gal (liquid)	8.810 l
0.0881 hl	16 pt (dry)
537.605 in.^{3}	8 qt (dry)

pennyweight =

0.4 dr (apoth/troy)	0.0034286 lb (avdp)
0.8777 dr (avdp)	1555 mg
1.555 g	0.05 oz (apoth/troy)
24 gr	0.05486 oz (avdp)
0.001555 kg	1.20 scruples (apoth)
0.0041667 lb (apoth/troy)	

percent efficiency: EFFICIENCY, PERCENT

percent grade: measure of slope in unit rise per 100 units distance, as 1 ft/100 ft.

perch: measure of length or measure of area.

perch: ROD

perch (masonry) =

     1 ft × 1.5 ft × 16.5 ft
     24.75 ft^3

perfect gas: GAS IDEAL

perfect vacuum: VACUUM, PERFECT

periodic table: chart of the elements in numerical order.

permittivity: ELECTROSTATICS

petroleum ether: benzine

pH: term used in the measure of acidity or alkalinity of a solution by the concentration of hydrogen ions in the solution; calculated as the logarithm of the reciprocal of the number of hydrogen ions per liter of solution, as

     $pH = \log 1/H^+$, where
     H^+ = number of hydrogen ions per liter of solution

pH scale: numerical scale for identifying the strength of acidity or alkalinity of a solution, ranging from 0 as most acidic to 7 for neutral and to 14 for most alkaline solutions.

phot: LAMBERT

pi (π): dimensionless constant equal to 3.14159.

pica (printer's) =

0.42175 cm   0.1660 in.

pico (p): prefix, equals 1 E − 12

pint (dry) =

0.015625 bushel (level)   0.1454 gal (liquid)
551 cm^3 33.60 in.^{3}
0.55 dal   0.5506 l
0.019445 ft^3 0.0625 peck
0.125 gal (dry)   0.5 qt (dry)

pint (liquid) =

0.01343 bushel (level)   0.4732 l
473.2 cm^3 4.732 E − 04 m^3
128 dr (liquid)   7680 minims
0.01671 ft^3 473.2 ml
0.125 gal (liquid)   16 oz (liquid)
4 gills   0.5 qt (liquid)
28.875 in.^{3} 6.189 E − 04 yd^3

piping formulas: FLUID FLOW

pitot tube: small-diameter tube bent at 90°, used to measure total pressure of a fluid or gas flowing in a pipe (or duct). A combination pitot/static tube is used in measuring total and static pressures simultaneously to determine velocity pressure.

Velocity pressure = total pressure − static pressure

Planck's constant =

6.625 E − 27 erg-s   6.625 E − 34 J-s
1.5835 E − 34 g-cal-s   3.9905 E − 10 J-s/Avogadro's number

Planck's law: HEAT TRANSFER
Pm: promethium (element)
Po: polonium (element)
poise: metric unit of absolute (dynamic) viscosity of a fluid.
poise =

100 cP	1 E+06 μP
1 dyn-s/cm^2	100 mPa-s
1 g/cm/s	0.10 N-s/m^2
0.0010197 g-s/cm^2	0.100 Pa-s
0.10 kg/m/s	0.0672 pdl-s/ft^2
0.010197 kg-s/m^2	1/rhe
0.002089 lb (force)-s/ft^2	0.002089 slug/ft/s
1.45 E−05 lb (force)-s/in.^{2}	stoke × specific gravity
0.0672 lb (mass)/ft/s	

Poiseuille's law: FLUID FLOW
Poisson's ratio: STRENGTH OF MATERIALS
polar moment of inertia: MOMENT OF INERTIA, PLANE AREA
polar section modulus: STRENGTH OF MATERIALS
polygon: geometric plane figure enclosed by three or more straight sides.
polygon, regular: polygon with all sides equal, where

a, b, c, d = dimensions, in.
    A = area, in.^{2}
    cg = center of gravity
    n = number of sides

area,

$$A = d^2 n \tan (180/n)$$
$$c = a/[2 \sin (180/n)]$$

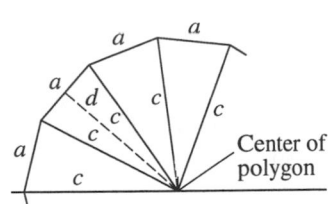

center of gravity, at center of polygon,

$$d = a/[2 \tan (180/n)]$$

polytropic process: THERMODYNAMICS
pond: GRAM
potential: ELECTRICAL

potential energy: stored energy of a body due to its elevation above a reference point (datum), where

> E = potential energy, ft-lb
> h = height of center of gravity of the body above datum, ft
> W = weight of the body, lb

potential energy,

$$E = Wh$$

potential energy head: FLUID FLOW

pound: lb; English unit for weight or force in gravitational system of units or mass in absolute system of units.

pound: pound (avdp)

pound =

0.01 cental	0.045359 myriagrams
116.667 dr (apoth/troy)	4.448 N
256 dr (avdp)	14.5833 oz (apoth/troy)
4.448 E+05 dyn	16 oz (avdp)
453.59 g	32.174 pdl
7000 gr	453.59 ponds
4.5359 hg	291.667 pwt (troy)
0.01 hwt	0.004536 quintal
0.04448 J/cm	350 scruples (apoth)
4.448 J/m	0.03108 slug
0.45359 kg	5 E−04 ton
1.21528 lb (apoth/troy)	4.464 E−04 ton (long)
453,592 mg	4.536 E−04 ton (metric)

pound/circular mil-foot: POUND/MIL/FOOT

pound/cubic foot =

0.01602 g/cm^3	0.13368 lb/gal (liquid)
16.02 g/l	5.787 E−04 lb/in.^{3}
0.01602 g/ml	5.456 E−09 lb/mil/ft
1.602 kg/hl	27 lb/yd^3
16.018 kg/m^3	16.018 mg/l
1.2445 lb/bushel	16 oz/ft^3

pound/cubic foot — *continued*

0.00926 oz/in.^{3} 0.0135 ton/yd^3
5 E − 04 ton/ft^3

pound/cubic inch =

27.68 g/cm^3 9.425 E − 06 lb/mil/ft
27,682 g/l   46,656 lb/yd^3
27.68 g/ml   2.7648 E + 04 oz/ft^3
2768 kg/hl   16 oz/in.^{3}
2.768 E + 04 kg/m^3 0.864 ton/ft^3
1728 lb/ft^3 23.328 ton/yd^3
231 lb/gal

pound/cubic yard =

5.934 E − 04 g/cm^3 2.143 E − 05 lb/in.^{3}
0.059327 kg/hl   0.59259 oz/ft^3
0.59327 kg/m^3 3.429 E − 04 oz/in.^{3}
0.04609 lb/bushel   1.852 E − 05 ton/ft^3
0.03704 lb/ft^3 5.0 E − 04 ton/yd^3
0.00495 lb/gal

pound/foot =

14.882 g/cm   14.59 N/m
37.80 g/in.   2.64 ton/mi
1.488 kg/m   2.357 ton (long)/mi
0.0833 lb/in.   1.488 ton (metric)/km
3.2808 lb/m

pound/gallon (dry) = 0.103 g/cm^3
pound/gallon (liquid) =

0.1198 g/cm^3 201.97 lb/yd^3
119.83 g/l   119.69 oz/ft^3
0.1198 g/ml   0.06926 oz/in.^{3}
119.83 kg/m^3 1.19832 E + 05 ppm
7.48 lb/ft^3 0.00374 ton/ft^3
0.00433 lb/in.^{3} 0.10099 ton/yd^3

pound/hour =

0.45359 kg/h	2.778 E − 04 lb/s
1.26 E − 04 kg/s	5 E − 04 ton/h
0.0167 lb/min	8.333 E − 06 ton/min

pound/inch =

178.579 g/cm	175.13 N/m
5443 g/ft	6.2992 oz/cm
453.6 g/in.	16 oz/in.
17.8579 kg/m	31.68 ton/mi
12 lb/ft	28.2857 ton (long)/mi
39.37 lb/m	17.857 ton (metric)/km
1.78579 E + 05 mg/cm	

pound/meter =

0.3048 lb/ft
0.0254 lb/in.

pound/mil/foot: pound/circular mil-foot
pound/mil/foot =

2.937 E + 06 g/cm^3	1.8335 E + 08 lb/ft^3
2.937 E + 09 kg/m^3	1.061 E + 05 lb/in.^{3}

pound/minute =

27.22 kg/h	0.0167 lb/s
0.4536 kg/min	0.03 ton/h
0.00756 kg/s	5 E − 04 ton/min
60 lb/h	

pound/second =

0.4536 kg/s	1.8 ton/h
3600 lb/h	0.03 ton/min
60 lb/min	

pound/square foot =

4.725 E − 04 atm	4.882 kg/m^2
4.788 E − 04 bar	0.04788 kPa
478.8 baryes	0.00694 lb/in.^{2}
0.03591 cm (mercury, 0°C, 32°F)	0.4788 mbar
478.8 dyn/cm^2	0.3591 mm (mercury, 0°C, 32°F)
0.01602 ft (water, 4°C, 39.2°F)	4.882 mm (water, 4°C, 39.2°F)
0.4882 g/cm	47.88 N/m^2
0.01439 in. (mercury, 0°C, 32°F)	0.1108 oz/in.^{2}
0.19221 in. (water, 4°C, 39.2°F)	47.88 Pa
4.882 E − 04 kg/cm^2	0.35913 torr

pound/square inch =

0.068046 atm	144 lb/ft^2
0.068948 bar	0.051715 m (mercury, 0°C, 32°F)
6.8947 E + 04 baryes	68.9476 mbar
5.1715 cm (mercury, 0°C, 32°F)	51,715 μm (mercury, °C, 32°F)
70.305 cm (water, 4°C, 39.2°F)	51.715 mm (mercury, 0°C, 32°F)
6.8947 E + 04 dyn/cm^2	703.05 mm (water, 4°C, 39.2°F)
0.1697 ft (mercury, 0°C, 32°F)	0.0068947 MPa
2.306 ft (water, 4°C, 39.2°F)	6894.73 N/m^2
70.307 g/cm^2	2304 oz/ft^2
2.036 in. (mercury, 0°C, 32°F)	16 oz/in.^{2}
27.679 in. (water, 4°C, 39.2°F)	6894.73 Pa
0.07031 kg/cm^2	0.072 ton/ft^2
703.07 kg/m^2	5 E − 04 ton/in.^{2}
7.0307 E − 04 kg/mm^2	51.715 torr
6.8947 kPa	

pound-foot: FOOT-POUND

pound-foot-second^2: slug-ft^2; English unit for moment of inertia of solids in terms of mass.

pound-inch: INCH-POUND

pound-inch-second^2: English units for moment of inertia of solids in terms of mass.

pound-square foot: English unit for moment of inertia of solids in terms of weight.

pound-square foot =

4.214 E + 05 g-cm^2	0.04214 kg-m^2
421.4 kg-cm^2	0.03108 slug-ft^2

pound-square inch: English unit for moment of inertia of solids in terms of weight.

pound-square inch =

2926.4 g-cm^2	0.00695 lb-ft^2
2.9264 kg-cm^2	2.158 E − 04 slug-ft^2

pound (apoth): POUND (APOTH)

pound (apoth) =

96 dr (apoth/troy)	3.73241 E + 05 mg
210.65 dr (avdp)	12 oz (apoth/troy)
373.2 g	13.166 oz (avdp)
5760 gr	240 pwt
3.7324 hg	288 scruples (apoth)
0.3732 kg	4.1143 E − 04 ton
0.822857 lb (avdp)	3.6735 E − 04 ton (long)
1 lb (troy)	3.7324 E − 04 ton (metric)

pound (avdp); pound (avoirdupois): POUND.

pound (force): English term for force in gravitational system of units.

pound (force) =

4.448 E + 05 dyn	4.448 N
0.453594 kg	32.174 pdl

pound (force)-second/square foot: English unit for absolute (dynamic) viscosity of a fluid.

pound (force)-second/square foot: SLUG/FOOT/SECOND

pound (force)-second/square inch: English unit for absolute (dynamic) viscosity of a fluid.

pound (force)-second/square inch: REYNE

pound (mass): English term for mass in absolute system of units.
pound (mass) =

453.6 g	0.0311 slug
0.45359 kg	

pound (mass)/foot/hour: English unit for absolute (dynamic) viscosity of a fluid.
pound (mass)/foot/hour =

0.41336 cP	0.000413 N-s/m^2
0.0041338 dyn-s/cm^2	0.004134 P
0.0041338 g/cm/s	0.0002778 pdl-s/ft^2
4.13338 E$-$04 kg/m/s	0.03108 slug/ft/h
8.63 E$-$06 lb (force)-s/ft^2	8.6336 E$-$06 slug/ft/s
0.0002778 lb (mass)/ft/s	

pound (mass)/foot/second: English unit for absolute (dynamic) viscosity of a fluid.
pound (mass)/foot/second =

1488.2 cP	1.488 E$+$07 μP
14.882 dyn-s/cm^2	1488 mPa-s
14.882 g/cm/s	1.4882 N-s/m^2
1.4882 kg/m/s	14.882 P
0.1516 kg-s/m^2	1.4882 Pa-s
0.0311 lb (force)-s/ft^2	1 pdl-s/ft^2
2.159 E$-$04 lb (force)-s/in.^{2}	0.0311 slug/ft/s

pound (mass)/inch/second: English unit for absolute (dynamic) viscosity of a fluid.
pound (mass)/inch/second =

1.7858 E$+$04 cP	1.785 E$+$08 μP
178.58 dyn-s/cm^2	1.7858 E$+$04 mPa-s
178.58 g/cm/s	0.37297 lb (force)-s/ft^2
17.858 kg/m/s	0.00259 lb (force)-s/in.^{2}
1.819 kg-s/m^2	12 lb (mass)/ft/s

pound (mass)/inch/second — *continued*

17.858 N-s/m^2	12 pdl-s/ft^2
178.58 P	0.37297 slug/ft/s
17.858 Pa-s	

pound (troy): POUND (APOTH)
pound (weight): POUND
pound/square inch (absolute pressure); psia:

$$psia = psig + 14.696$$

pound/square inch (gauge pressure); psig: pressure as indicated on the pressure gauge.

poundal: English unit of force in absolute system of units; 1 poundal of force imparts a uniform acceleration of 1 ft/s^2 upon a mass of 1 lb.

poundal =

1.3826 E + 04 dyn	0.1383 J/m
14.098 g	0.0141 kg
0.14098 hg	0.03108 lb (avdp)
0.001383 J/cm	0.1383 N

poundal/cubic inch =

843.714 dyn/cm^3	0.86032 g/cm^3

poundal/inch =

5443 dyn/cm	5.5504 g/cm

poundal/square foot = 1.488 Pa
poundal/square inch =

2143 dyne/cm^2	2.1852 g/cm^2

poundal-second/square foot: English unit of absolute (dynamic) viscosity of a fluid.
poundal-second/square foot: POUND (MASS)/FOOT/SECOND
power: rate at which work is being done in work per unit time, as ft-lb/min, hp, etc.

power factor: ELECTRICAL

power of a number, exponent: for example,

a^n is number a multiplied by itself n times.
a^{-n} is the reciprocal of number a multiplied by itself n times.

Pr: praseodymium (element)

Prandtl number (N_{PR}): dimensionless number used in heat transfer calculations for fluid flow, where

C_P = specific heat of fluid, Btu/lb (mass)/°F
k = coefficient of heat conductivity of the fluid, Btu/h-ft-°F
μ = absolute viscosity of the fluid, lb (mass)/ft/h

Prandtl number, $N_{PR} = \mu C_P / k$

pressure: measurement of force per unit area, lb/ft^2, lb/in.^{2}, g/cm^2, etc.

pressure, absolute: ABSOLUTE PRESSURE

pressure, gauge: GAUGE PRESSURE

pressure drop: term used in piping calculations denoting the loss of pressure in the flow of liquids (or gases) through a piping system as a result of pipe friction, fittings, etc.

pressure drop in piping: AIR FLOW; FLUID FLOW; GAS FLOW; STEAM FLOW.

pressure energy: term used in pump calculations for total energy required for the flow of a fluid, in ft-lb/lb.

prism: CUBE, RECTANGULAR PRISM

producer's gas: a by-product resulting from the addition of air to coke gases, containing approximately

50.9% nitrogen	4.5% carbon dioxide
27.0% carbon monoxide	3.0% methane
14.0% hydrogen	0.6% oxygen

progression, arithmetic: ARITHMETIC PROGRESSION

progression, geometric: GEOMETRIC PROGRESSION

projectile: a body projected by a force.

projectile path: TRAJECTORY FORMULAS

Prony brake: a test device used to determine torque requirement at a shaft or pulley of machinery. It may be done manually with a clamp on the shaft or pulley with an extended lever. The force × lever required to rotate the shaft is the torque required, where

F = force applied, lb
hp = horsepower
L = length of lever arm, ft
N = speed, rpm
T = torque, ft-lb

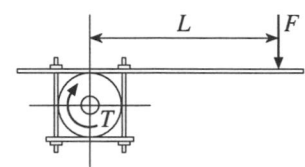

horsepower, hp = 1.904 E−04 FLN
torque, $T = FL$

proportional limit: STRENGTH OF MATERIALS
proton: positive charge in the nucleus of an atom.
proton mass =

   1.67 E−24 g   1.67 E−27 kg

psychrometer: the instrument used in determining the dry and wet bulb temperatures of air simultaneously. The sling type psychrometer consists of two (2) thermometers on a frame that is rotated rapidly in the air that is to be measured. One thermometer is equipped with a piece of gauze attached to the thermometer bulb. The gauze is moistened with water to determine the wet bulb temperature of the air while the other thermometer reads the dry bulb temperature. Psychrometer readings may then be used to determine other properties of the air by referring to the Psychrometric Chart.

Psychrometric Chart: chart indicating properties of air and water-vapor mixtures, including dry bulb temperature, wet bulb temperature, dew point temperature, water vapor per pound of air, relative humidity, enthalpy and vapor pressure. Charts are published by ASHAE, various equipment manufacturers, and can be found in handbooks.

Pt: platinum (element)
Pu: plutonium (element)

PULLEY FORMULAS, Drives: where

c = distance between centers, in.
D_1 = diameter of small pulley, in.
D_2 = diameter of large pulley, in.
L = length of belt, in.
N_1 = speed of small pulley, rev/min
N_2 = speed of large pulley, rev/min
V = belt speed, ft/min

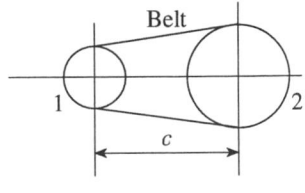

belt length,

$$L = 2c + 1.57(D_1 + D_2) + 0.25(D_2 - D_1)^2/c$$

belt speed,

$$V = 0.2618D_1N_1 = 0.2618D_2N_2$$

·····" ·y size,

$$D_1 = D_2N_2/N_1$$
$$D_2 = D_1N_1/N_2$$

PULLEY FORMULAS, lifting advantage: where

D = diameter of pulley, in.
F = force, lb
n = number of strands
W = weight of load, lb

force required, two pulleys on same shaft,

$$F = WD_1/D_2$$

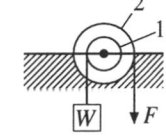

force required, one pulley fixed, one pulley floating,

$$F = W/2$$

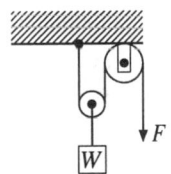

force required, two pulleys with multiple strands, as with a block-and-tackle system,

$$F = W/n$$

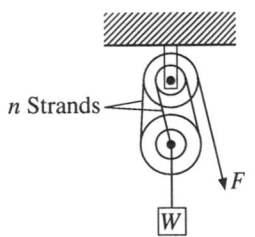

PUMP FORMULAS FOR FLUIDS: where

bhp = brake horsepower
C = pump displacement, in.^{3}/rev
DH = discharge head, ft (fluid)
eff = efficiency, decimal
G = flow of fluid, gal/min
h = static head, ft (fluid)
h_F = head loss, friction, ft (fluid)
h_{SD} = static discharge head, ft (fluid)
h_{SL} = static suction lift head, ft (fluid)
h_{SS} = static suction head, ft (fluid)
h_{ST} = total static head, ft (fluid)
h_V = velocity pressure head, ft (fluid)
H = total head at pump, ft (fluid)
hp = pump horsepower
KE = kinetic energy, ft-lb
N = pump speed, rev/min
p = pressure of liquid, psia
Q = pump capacity, gal/min
S = specific gravity of fluid
SH = suction head, ft (fluid)
SL = suction lift, ft (fluid)
T = torque, in.-lb
V = velocity of fluid, ft/sec
w = specific weight of fluid, lb/ft^3
W = weight flow of fluid, lb/min

brake horsepower: horsepower input required at pump,

$$\text{bhp} = \text{hp/eff}$$

capacity:

$$Q = CN/231$$

discharge head (DH): energy at discharge of pump in operation,

$$\text{DH} = h_{SD} + h_F + h_V \text{ (discharge side of pump)}$$

efficiency:

$$\text{eff} = \text{hp (output)/hp (input)}$$

PUMP FORMULAS FOR FLUIDS—*continued*

flow:

$$Q = 0.12W/S$$
$$W = 8.34QS$$

friction, head loss (h_F): loss in pressure due to flow resistance of piping, fittings, etc.; see FLUID FLOW

head pressure conversion to psia:

$$p = 0.0069wh$$
$$= 0.4322h/S$$

horsepower input required: brake horsepower *
horsepower output:

$$\text{hp} = WH/33,000$$
$$= GHS/3960$$
$$= TN/63,025$$

kinetic energy: energy due to motion of a fluid,

$$\text{KE} = 0.0155V^2W$$

pressure conversion to head:

$$h = 144p/w = 2.31p/S$$

speed of pump:

$$N = 231Q/C$$

static head (h): pressure due to height when a liquid is at rest; four types:

1. static discharge head (h_{SD}), the vertical distance from the centerline of a pump to the discharge level.
2. static suction head (h_{SS}), the vertical distance from the supply level to the centerline of a pump; the source of supply is above the discharge outlet of the pump.
3. static suction lift head (h_{SL}), the vertical distance from the supply level to the centerline of the pump; the source of supply is below the pump.
4. total static head (h_{ST}), the vertical distance from the supply level to the discharge level,

$$h_{ST} = h_{SD} + h_{SL}$$
$$= h_{SD} - h_{SS}$$

PUMP FORMULAS FOR FLUIDS — *continued*

suction head (SH): energy at the intake of a pump in operation and the pressure is above atmospheric; the source of supply is above the centerline of pump.

$$\text{SH} = h_{SS} - h_F - h_V \text{ (suction side of pump)}$$

suction lift (SL): energy at the intake of a pump in operation; the pressure is below atmospheric; the source of supply is below the centerline of the pump.

$$\text{SL} = h_{SL} + h_F + h_V \text{ (suction side of pump)}$$

torque:

$$T = 0.159CH$$
$$= 36.76QH/N$$

total head (H): pump differential pressure head; pressure difference between discharge and intake of pump in operation.

$$H = \text{DH} - \text{SH}$$
$$= \text{DH} + \text{SL}$$

velocity pressure head: energy of a fluid due to its velocity,

$$h_V = 0.0155V^2$$

PYRAMID, regular: polygon base and the sides meet at the vertex, where

A = area
b = length of base side
c = slant height
cg = center of gravity
d = perpendicular distance from center of polygon base to side
h = height of pyramid
n = number of sides
s = perimeter of base

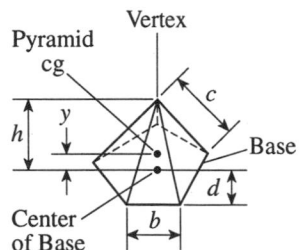

PYRAMID, regular — *continued*

area of base,

$$A = 0.5bdn$$

area of one side,

$$A = 0.5bc$$

area of sides, total,

$$A = 0.5bcn = 0.5cs$$

center of gravity (cg), on line from cg of base to vertex, at

$$y = 0.25h$$

distance from center of polygon to side,

$$d = 0.5b/\tan(180/n)$$

perimeter of base,

$$s = bn$$

volume,

$$V = 0.1667bdhn$$

PYRAMID, regular: rectangular base with opposite sides equal, where

a, b = base sides $(a > b)$, ft
A = area, ft^2
c = slant height, ft
cg = center of gravity
h, y = dimensions, ft
I_{AA}, I_{BB} = moment of inertia, slug-ft^2
k_{AA}, k_{BB} = radius of gyration, ft
V = volume, ft^3
W = weight of pyramid, lb

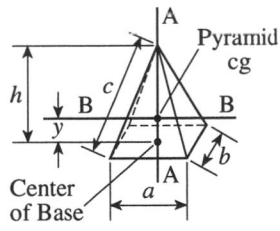

area of total surface,

$$A = ab + ac + bc$$

PYRAMID, regular — *continued*

center of gravity (cg), through center of pyramid, at

$$y = 0.25h$$

moment of inertia,

$$I_{AA} = 0.00155W(a^2 + b^2)$$
$$I_{BB} = 0.00155W(b^2 + 0.75h^2)$$

radius of gyration,

$$k_{AA} = 0.2236(a^2 + b^2)^{1/2}$$
$$k_{BB} = 0.2236(b^2 + 0.75h^2)^{1/2}$$

volume,

$$V = 0.333abh$$

Q

quadrant (of circle) =

0.25 circle	1.571 rad
90°	0.25 rev
5400 min	3.24 E+05 s

quadratic equation: $Ax^2 + Bx + C = 0$, where A, B, and C are constants, solution for x,

$$x = [-B(\pm)(B^2 - 4AC)^{1/2}]/2A$$

quantum = 6.624 E−27 erg/s

quart (dry) =

0.009524 bbl (dry)	0.0011 kl
0.03125 bushel (level)	1.10122 l
1101 cm³	0.0011 m³
1.10122 dm³	0.125 peck
0.03889 ft³	2 pt (dry)
0.25 gal (dry)	1.16365 qt (liquid)
0.24235 gal (Imperial)	0.0011 stere
0.2909 gal (liquid)	0.00144 yd³
67.20 in.³	

quart (liquid) =

0.026855 bushel (level)	0.94635 dm³
946.4 cm³	256 dr (liquid)

quart (liquid) — *continued*

0.03342 ft³	15,360 minims
0.2149 gal (dry)	946.4 ml
0.20828 gal (Imperial)	32 oz (liquid)
0.25 gal (liquid)	2 pt (liquid)
8 gills	0.8594 qt (dry)
57.75 in.³	0.833 qt (Imperial)
9.4636 E−04 kl	9.4636 E−04 stere
0.9463 1	0.001238 yd³
9.464E−04 m³	

quintal: hundredweight in the metric system

quintal =

1 E+05 g	220.46 lb (avdp)
100 kg	

quire =

0.05 ream of paper
25 sheets of paper

R

Ra: radium (element)
radial acceleration: ROTATION
radial velocity: ROTATION
radian: the angle in a circle where the radius of the arc equals the
length of the arc, specifically 57° and 17.75 min.
radian =

0.159155 circle	0.6366 quadrant
57.29578° (angular)	0.159155 rev
3438 min (angular)	2.063 E + 05 sec (angular)

radian/second =

3437.75° (angular)/min	9.549 rev/min
57.29578° (angular)/s	0.1592 rev/s

radian/second/second =

573 rev/min^2	0.1592 rev/s^2
9.549 rev/min/s	

radiant heating: HEAT TRANSFER
radiation factor: HEAT TRANSFER
radius of gyration, plane area: as related to the moment of inertia of the
plane surface about a referenced axis. It is the equivalent of concentrating the entire area of the plane surface at a point a distance k

radius of gyration, plane area — *continued*

from the referenced axis, where

A = area, in.^{2}
I = moment of inertia, in.^{4}
k = radius of gyration, in.

⊗ radius of gyration:

$$k = (I/A)^{1/2}$$

radius of gyration for various sections: refer to ANGLE; CHANNEL; CIRCLE; ELLIPSE; HEXAGON; OCTAGON; PARABOLA; RECTANGLE; SQUARE; TEE; TRAPEZOID; TRIANGLE; Z BAR.

radius of gyration, solid: as related to the moment of inertia of the solid, in units of mass, about a referenced axis. It is the equivalent of concentrating the mass of the body a distance k from the referenced axis, where

I = moment of inertia, slug-ft^2
k = radius of gyration, ft
m = mass, slugs
W = weight, lb

⊗ radius of gyration:

$$k = (I/m)^{1/2}$$
$$= 5.67(I/W)^{1/2}$$

radius of gyration for various solids: refer to CONE; CUBE; CYLINDER; DISK; ELLIPSOID; PYRAMID; RECTANGULAR PRISM; RING; ROD; SPHERE.

radius of gyration (angular): ROTATION
Ramden's chain: CHAIN, RAMDEN
Ramden's link: FOOT
Rankine: term used for absolute Fahrenheit temperature values.
Rb: rubidium (element)
Re: rhenium (element)
reactance: ELECTRICAL
reaction: STATICS

ream: measure of quantity of paper
ream =

    20 quires
    500 sheets

rectangle: four-sided figure with opposite sides equal

    area,

$$A = ab$$

    diagonal,

$$d = (a^2 + b^2)^{1/2}$$

    perimeter,

$$s = 2ab$$

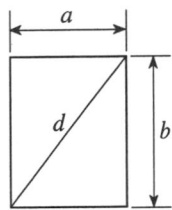

RECTANGLE SECTION FORMULAS: where

$$a, b, x, y, z = \text{dimensions, in.}$$
$$A = \text{area, in.}^2$$
$$cg = \text{center of gravity}$$
$$I_{AA}, I_{BB}, I_{CC}, I_{DD} = \text{moment of inertia, in.}^4$$
$$I_P = \text{polar moment of inertia, in.}^4$$
$$k_{AA}, k_{BB}, k_{CC}, k_{DD} = \text{radius of gyration, in.}$$
$$k_P = \text{polar radius of gyration, in.}$$
$$Z_{AA}, Z_{BB}, Z_{CC}, Z_{DD} = \text{section modulus, in.}^3$$
$$Z_P = \text{polar section modulus, in.}^3$$

rectangle, hollow:

    area,

$$A = ab - cd$$

    center of gravity (cg),

$$x = 0.5a$$
$$y = 0.5b$$

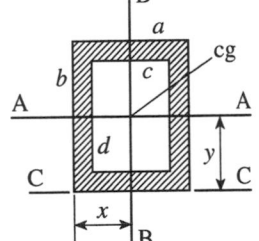

RECTANGLE SECTION FORMULAS — *continued*

moment of inertia,

$$I_{AA} = 0.0833(b^3a - d^3c)$$
$$I_{BB} = 0.0833(a^3b - c^3d)$$
$$I_{CC} = 0.3333b^3a - 0.0833cd(3b^2 + d^2)$$
$$I_P = 0.0833[ab(a^2 + b^2) - cd(c^2 + d^2)]$$

radius of gyration,

$$k_{AA} = \frac{0.2887(b^3a - d^3c)^{1/2}}{(ab - cd)^{1/2}}$$

$$k_{BB} = \left[\frac{0.0833(a^3b - c^3d)}{ab - cd}\right]^{1/2}$$

$$k_{CC} = \frac{[0.3333b^3a - 0.0833cd(3b^2 + d^2)]^{1/2}}{(ab + cd)^{1/2}}$$

section modulus,

$$Z_{AA} = 0.1667(b^3a - d^3c)/b$$
$$Z_{BB} = 0.1667(a^3b - c^3d)/a$$
$$Z_{CC} = 0.3333b^2a - 0.0833cd(3b + d^2)$$

rectangle, solid:

area,

$$A = ab$$

center of gravity (cg),

$$x = 0.5a$$
$$y = 0.5b$$
$$z = ab/(a^2 + b^2)^{1/2}$$

moment of inertia,

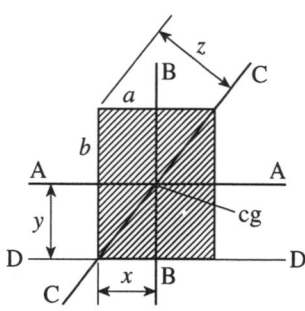

$$I_{AA} = 0.0833b^3a$$
$$I_{BB} = 0.0833a^3b$$
$$I_{CC} = (0.1667a^3b^3)/(a^2 + b^2)$$
$$I_{DD} = 0.3333b^3a$$
$$I_P = 0.0833ab(a^2 + b^2)$$

RECTANGLE SECTION FORMULAS — *continued*

radius of gyration,

$$k_{AA} = 0.2887b$$
$$k_{BB} = 0.2887a$$
$$k_{CC} = (0.4082ab)/(a^2 + b^2)^{1/2}$$
$$k_{DD} = 0.5774b$$
$$k_P = 0.2887(a^2 + b^2)^{1/2}$$

section modulus,

$$Z_{AA} = 0.1667b^2a$$
$$Z_{BB} = 0.1667a^2b$$
$$Z_{CC} = (0.1667a^2b^2)/(a^2 + b^2)^{1/2}$$
$$Z_{DD} = 0.333b^2a$$
$$Z_P = a^2b/(3 + 1.8a/b)$$

RECTANGULAR PRISM (BLOCK) FORMULAS: where

a, b, c, x, y, z = dimensions, ft
A = area of surface, ft^2
cg = center of gravity
I_{AA}, I_{BB}, I_{CC} = moment of inertia, slug-ft^2
k_{AA}, k_{BB}, k_{CC} = radius of gyration, ft
V = volume, ft^3
W = weight, lb

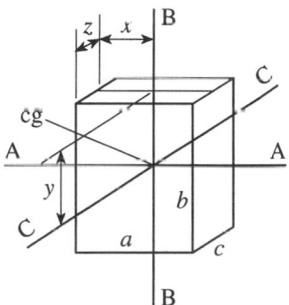

area of surface,

$$A = 2(ab + ac + bc)$$

center of gravity (cg),

$$x = 0.05a$$
$$y = 0.5b$$
$$z = 0.5c$$

moment of inertia,

$$I_{AA} = 0.00259W(b^2 + c^2)$$
$$I_{BB} = 0.00259W(a^2 + c^2)$$
$$I_{CC} = 0.00259W(a^2 + b^2)$$

RECTANGLE PRISM (BLOCK) FORMULAS — *continued*

radius of gyration,

$$k_{AA} = 0.2887(b^2 + c^2)^{1/2}$$
$$k_{BB} = 0.2887(a^2 + c^2)^{1/2}$$
$$k_{CC} = 0.2887(a^2 + b^2)^{1/2}$$

volume,

$$V = abc$$

---

rectilinear motion: LINEAR MOTION
refrigeration cycle sequence:

1. **compressor:** increases pressure of the saturated vapor, resulting in a superheated vapor.
2. **condenser:** heat is transferred from the superheated vapor to an air or water medium resulting in a high-pressure saturated liquid.
3. **expansion valve:** reduces the high-pressure liquid to a lower-pressure liquid.
4. **evaporator:** heat is absorbed by the refrigerant, changing it from a low-pressure liquid to a low-pressure saturated vapor and returns it to the compressor.

REFRIGERATION FORMULAS: where

c = piston clearance, decimal
C = cylinder volume, in.^{3}
d = piston displacement, in.^{3}
D = piston displacement per revolution, in.^{3}/rev
eff = efficiency, decimal
h_1 = enthalpy, evaporator "in," Btu/lb
h_2 = enthalpy, evaporator "out," compressor "in," Btu/lb
h_3 = enthalpy, compressor "out," condenser "in," Btu/lb
h_4 = enthalpy, condenser "out," Btu/lb
H_C = condenser heat transferred, Btu/lb
H_E = evaporator heat transferred, Btu/lb
hp = compressor horsepower
N = compressor speed, rev/min
P = compressor power, ft-lb/min
Q = compressor capacity, ft^3/min

REFRIGERATION FORMULAS — *continued*

Q_H = heat flow, Btu/min
Q_R = refrigerant flow rate, lb/min
R = refrigeration rate, ton
T = ton, refrigeration
w = specific weight of refrigerant, lb/ft^3
W = compressor work, Btu/lb

clearance (c): clearance above piston at top of stroke,

$$c = 1 - d/C$$

coefficient of performance: efficiency *
compressor:

capacity,

$$Q = Q_R/w = ND/1728$$

horsepower,

$$\text{hp} = P/33{,}000 = 4.715RW/H_E$$
$$= 4.715R(h_3 - h_2)/(h_2 - h_1)$$

power,

$$P = 778wWQ$$
$$= 778Q_R(h_3 - h_2)$$

work,

$$W = h_3 - h_2$$

condenser, heat transferred: $h_C = h_3 - h_4$
cylinder volume (C): volume above piston with the piston at the bottom of the stroke.
efficiency: coefficient of performance; ratio of the change of enthalpy in the evaporator to the change of enthalpy through the compressor,

$$\text{eff} = H_E/W = (h_2 - h_1)/(h_3 - h_2)$$

evaporator, heat transferred: $H_E = h_2 - h_1$
heat flow:

$$Q_H = Q_R H_E$$
$$= Q_R(h_2 - h_1)$$

REFRIGERATION FORMULAS — *continued*

piston displacement (d): cylinder volume displaced by piston travel.
piston displacement per revolution:

$$D = d/\text{rev}$$

refrigerant flow rate:

$$Q_R = wQ$$
$$= Q_H/H_E$$

refrigeration rate:

$$R = Q_R H_E/200$$
$$= Q_R(h_2 - h_1)/200$$

ton (refrigeration): equivalent to melting 1 ton (2000 lb) of ice at 32°F in a
24-h period,

$$T = 200 \text{ Btu/min}$$
$$= 12,000 \text{ Btu/h}$$

---

relative humidity: HUMIDITY, RELATIVE
reluctance: ELECTROMAGNETISM
resistance (electrical circuit): ELECTRICAL
resistivity (electrical) of various materials: ELECTRICAL
rev: revolution
revolution =

360° (angular)	4 quadrants
400 grades	6.283 rad

revolution/minute =

6° (angular)/s	0.01667 rev/s
0.1047 rad/s	

revolution/minute/minute =

0.001745 rad/s^2	2.778 E−04 rev/s^2
0.01667 rev/min/s	

revolution/minute/second =

 0.1047 rad/s^2 0.01667 rev/s^2
 60 rev/min^2

revolution/second =

 21,600° (angular)/min   6.283 rad/s
 360° (angular)/s   60 rev/min

revolution/second/second =

 6.283 rad/s^2 60 rev/min/s
 3600 rev/min^2

reyne, reyn: English unit for absolute (dynamic) viscosity of a fluid.
reyne =

 6.89 E+06 cP   6.89 E+10 μP
 6.8947 E+04 dyn-s/cm^2 6.89 E+06 mPa-s
 6.8947 E+04 g/cm/s   6895 N-s/m^2
 6895 kg/m/s   6.8947 E+04 P
 702.7 kg-s/m^2 6895 Pa-s
 144 lb (force)-s/ft^2 4633 pdl-s/ft^2
 1 lb (force)-s/in.^{2} 144 slugs/ft/s
 4633 lb (mass)/ft/s

Reynold's number (N_R): AIR FLOW; FLUID FLOW; GAS FLOW; STEAM FLOW.
Rh: rhodium (element)
rhe: cm-s/g, cm^2/dyn/s; unit of fluidity of a liquid.
rhe =

 1 cm^2/dyn/s   10 m^2/N/s
 1 cm-s/g   1/P

rhomboid: four-sided plane figure with two sides parallel and adjacent
   sides equal.
RHOMBUS: four-sided plane figure with all four sides equal and
   opposite sides parallel,

RHOMBUS — *continued*

area,

$$A = ah = 0.5bc = 2a(\sin A)$$

diagonal,

$$b = 1.414a(1 - \cos A)^{1/2}$$
$$c = 1.414a(1 + \cos A)^{1/2}$$

height,

$$h = a(\sin A)$$

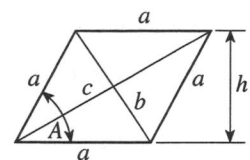

right angle: an angle equal to 90°.

right triangle: TRIANGLES, RIGHT

RING (TORUS): where

a = cross-sectional area, ft^2
A = surface area, ft^2
d = inside diameter, ft
D = mean diameter, ft
F = safe load on ring, lb
I_{AA}, I_{BB} = moment of inertia, slug-ft^2
k_{AA}, k_{BB} = radius of gyration, ft
S_W = working stress, lb/ft^2
t = cross-sectional diameter, ft
V = volume, ft^3
W = weight, lb

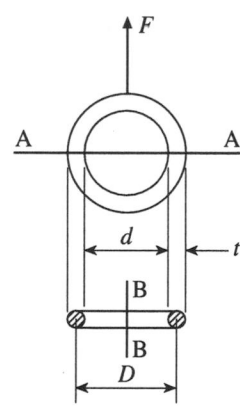

area, cross-section,

$$a = 0.7854t^2$$

area, surface,

$$A = 9.87tD$$

load on ring, safe,

$$F = 0.617t^3 S_W/D$$

mean diameter,

$$D = d + t$$

RING (TORUS)—*continued*

moment of inertia,

$$I_{AA} = 0.003885W(D^2 + 5t^2)$$
$$I_{BB} = 0.00777W(D^2 + 3t^2)$$

radius of gyration,

$$k_{AA} = 0.3536(D^2 + 5t^2)^{1/2}$$
$$k_{BB} = 0.5(D^2 + 3t^2)^{1/2}$$

stress, working,

$$S_W = 1.62FD/t^3$$

volume,

$$V = 2.4674Dt^2$$

Rn: radon (element)

Rockwell hardness: a reference number for hardness of materials. Numbers are based on the penetration depth when the penetrator is pressed into the material. System includes various penetrators corresponding to different scales.

rod, rood =

502.92 cm	0.00503 km
0.25 chain (Gunter)	25 links (Gunter)
0.165 chain (Ramden)	16.5 links (Ramden)
0.5029 dam	5.0292 m
16.5 ft	0.003125 mi (statute)
0.025 furlong	1 perch
0.0503 hm	5.5 yd
198 in.	

ROD, THIN: where

$$A = \text{cross-sectional area, ft}^2$$
$$\text{cg} = \text{center of gravity}$$
$$d, l, y = \text{dimensions; ft}$$
$$I_{AA}, I_{BB} = \text{moment of inertia, slug-ft}^2$$
$$k_{AA}, k_{BB} = \text{radius of gyration, ft}$$

ROD, THIN — *continued*

V = volume, ft^3
W = weight, lb

area, surface,

$$A = 1.5708d^2 + 3.1416dL$$

center of gravity (cg), through center of rod, at

$$y = 0.5L$$

moment of inertia,

$$I_{AA} = 0.00259L^2W$$
$$I_{BB} = 0.01036L^2W$$

radius of gyration,

$$k_{AA} = 0.2887L$$
$$k_{BB} = 0.5774L$$

volume,

$$V = 0.7854d^2L$$

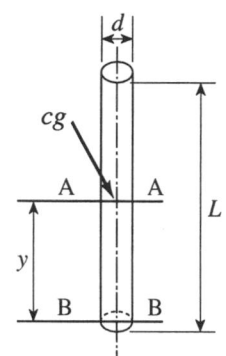

rolling friction coefficient: FRICTION
rood: ROD
root of a number: $a^{1/n}$ is the nth root of number a.
ROTATION: where

a = angular acceleration, rad/sec^2
a_N = normal acceleration, ft/sec^2
a_T = tangential acceleration, ft/sec^2
B = angular displacement, rad
B_1 = initial angle, rad
B_2 = final angle, rad
F_C = centrifugal force, lb
F_P = centripetal force, lb
F_T = tangential force, lb
g = gravitational acceleration, 32.174 ft/sec^2
hp = horsepower
I = moment of inertia about axis of rotation, lb-ft-sec^2 or slug-ft^2

ROTATION — *continued*

I_A = angular impulse, lb-ft-sec
k = radius of gyration, ft
KE = kinetic energy, ft-lb
m = mass (W/g), lb-sec^2/ft, slugs
M_A = angular momentum, lb-ft-sec
N = speed, rev/min
N_1 = initial speed, rev/min
N_2 = final speed, rev/min
r = radius, ft
t = time, sec
t_1 = initial time, sec
t_2 = final time, sec
T = torque, lb-ft
V_N = normal velocity, ft/sec
V_T = tangential velocity, ft/sec
w = angular velocity, rad/sec
w_1 = initial angular velocity, rad/sec
w_2 = final angular velocity, rad/sec
W = weight, lb
W_k = work, ft-lb

angle of movement: angular displacement ∗

angular acceleration: the continuous increase of the velocity of a body in a rotating motion,

$$a = w/t$$
$$= 0.1047N/t$$
$$= (w_2 - w_1)/(t_2 - t_1)$$
$$= 2(B_2 - w_1 t)/t^2$$

angular displacement: angle of movement; angle rotated through,

$$B_2 - B_1 = 0.5(w_1 + w_2)t$$
$$= w_1 t + 0.5t^2 a$$
$$= 0.5(w_2^2 - w_1^2)/a$$

angular impulse: torque imparted to a rotating body over a time interval to cause a change in the momentum of the body,

$$I_A = tT$$

ROTATION — *continued*

angular momentum: product of the moment of inertia and the angular velocity of a body rotating about its axis,

$$M_A = wI$$
$$= 0.1047IN$$
$$= 0.00325r^2NW$$

angular velocity: angular displacement of a body per unit time,

$$w = B/t$$
$$= 0.1047N$$
$$= (B_2 - B_1)/(t_2 - t_1)$$
$$= V_T/r$$

angular velocity at end of t seconds:

$$w_2 = w_1 + at$$
$$= (w_1 + 2aB)^{1/2}$$

center of percussion: point on a rotating body where the resultant of all the external forces passes through.

centrifugal force: force originating at the axis of a rotating body, acting radially and outward to the edge of the rotating body. It is the opposite action to that of centripetal force,

$$F_C = -F_P = -ma_N$$
$$= -0.0311V_T^2W/r$$
$$= -0.0311w^2Wr$$
$$= -3.41\,E-04\,N^2Wr$$

centripetal force: force acting radially, inward from the outer edge of a rotating body to the axis of the rotating body. It is the opposite action to the centrifugal force,

$$F_P = -F_C = ma_N$$
$$= 0.0311V_T^2W/r$$
$$= 0.0311w^2Wr$$

⊗ horsepower: work done by rotation of a body per unit time,

$$hp = 0.001818Tw$$
$$= 1.904\,E-04\,TN$$

ROTATION — *continued*

impulse, angular: angular impulse   *
kinetic energy:

$$KE = 0.5w^2 I$$
$$= 1.7 \, E - 04 \, k^2 N^2 W$$
$$= 0.0155k^2 w^2 W$$

moment: torque   *
moment of inertia (angular):

$$I = k^2 W/g$$
$$= 0.0311k^2 W$$

momentum, angular: angular momentum   *
normal acceleration: radial acceleration; acceleration toward the center of the circular path,

$$a_N = V_T^2/r$$
$$= w^2 r$$
$$= wV_T$$

normal velocity (V_N): radial velocity; velocity of a point where the distance from the point to the axis of rotation is changing, as in a cam action. In circular motion, the normal velocity is zero because the radius is constant.
radial acceleration: normal acceleration   *
radial velocity: normal velocity   *
radius of gyration (angular): distance from a point on a rotating body to the axis of rotation, where the equivalent mass of the body is concentrated on this point,

$$k = 5.67(I/W)^{1/2}$$

speed:

$$N = 9.55w$$
$$= 9.55V_T/r$$

tangential acceleration: rate of change of tangential velocity,

$$a_T = ra$$

ROTATION — *continued*

tangential force:

$$F_T = 0.00325kWN/t$$
$$= 0.00325k^2WN/tr$$

tangential velocity: instantaneous tangential velocity vector of a point on a rotating body,

$$V_T = rw$$
$$= 0.1047rN$$

time required to change speed of a rotating mass:

$$t = 0.003256k^2W(N_2 - N_1)/T$$
$$= 0.0311k^2W(w_2 - w_1)/T$$

torque: moment; twisting moment about an axis of rotation due to a force and its lever arm, evaluated as the product of the force and the perpendicular distance of the lever arm from the axis of rotation,

$$T = Fr$$
$$= 0.0311k^2Wa$$
$$= 0.5I(w_2^2 - w_1^2)/B$$
$$= I(w_2 - w_1)/t$$
$$= 0.1047(N_2 - N_1)I/t$$
$$= 0.00325k^2W(N_2 - N_1)/t$$
$$= 5252(hp)/N$$

velocity, angular: angular velocity   *
work: force moving through a distance, as the product of angular displacement and torque.

$$W_k = TB$$
$$= 0.1047TNt$$
$$= 3.4\,E-04k^2N^2W$$
$$= 0.0311k^2w^2W$$

---

rupture point: STRENGTH OF MATERIALS

S

S: sulfur (element)

saturated air: point where the dry and wet bulb temperatures of an air/ water-vapor mixture are equal; 100 percent humidity.

Saybolt Second Furol (SSF): American unit for kinematic viscosity of very heavy liquids. A larger orifice is used for measurement and the time of efflux is a tenth of that used in the Saybolt Second Universal system.

Saybolt Second Furol =

0.463 cSt
0.1 Saybolt Second Universal

Saybolt Second Universal (SSU); Saybolt Universal Second (SUS): American unit for kinematic viscosity unit of lighter liquids, by measuring the time required for 60 cm^3 of the liquid to flow through a standard orifice at a specified temperature.

Saybolt Second Universal =

4.63 cSt
10 Saybolt Second Furol

Saybolt Universal Second (SSU): SAYBOLT SECOND UNIVERSAL

Sb: antimony (element)

Sc: scandium (element)

score: 20 items

screen, screen openings: MESH
scruple (apoth) =

0.3333 dr (apoth/troy)   0.002857 lb (avdp)
0.73143 dr (avdp)   1296 mg
1.296 g   0.0417 oz (apoth/troy)
20 gr   0.0457 oz (avdp)
0.001296 kg   0.8333 pwt
0.003472 lb (apoth/troy)

Se: selenium (element)
secant: TRIGONOMETRY
second =

1.1574 E−05 d   0.01667 min
2.778 E−04 h

second (angular) =

2.778 E−04 degrees (angular)   4.848 E−06 rad
0.01667 min (angular)

section: measure of area, equal to 1 mi^2.
section modulus: STRENGTH OF MATERIALS
self-ignition of fuels: AUTOIGNITION
SHAFT FORMULAS: where

a = angle of twist, degrees
c = ratio of diameters of hollow shaft
d = inside diameter of hollow shaft, in.
D = outside diameter of shaft, in.
G = modulus of elasticity (shear), lb/in.^{2}
hp = horsepower
I_P = polar moment of inertia of shaft, in.^{4}
L = length of shaft, in.
N = speed, rev/min
r = radius of shaft, in.
sf = safety factor, code or design
S_U = shear strength, ultimate, lb/in.^{2}

SHAFT FORMULAS — *continued*

S_W = shear stress, working, lb/in.^{2}
t = thickness, in.
T = torque, in.-lb
Z_P = polar section modulus, in.^{3}

⊗ angle of twist (a): torsional deflection; total angle of twist for length of shaft.

hollow shaft,

$$a = 583.6TL/(D^4 - d^4)G$$

solid shaft,

$$a = 583.6TL/GD^4$$

diameter of hollow shaft:

$$D = 1.7205T^{1/3}/[S_W(1 - c^4)]^{1/3}$$

diameter of solid shaft:

$$D = 1.7205(T/S_W)^{1/3}$$

horsepower:

$$\text{hp} = 1.587 \text{ E}-05 \ TN$$

modulus of elasticity, shear (G): STRENGTH OF MATERIALS
polar moment of inertia:

hollow shaft,

$$0.09818(D^4 - d^4)$$

solid shaft,

$$0.09818D^4$$

polar section modulus:

$$Z_P = I_P/r$$

hollow shaft,

$$Z_P = 0.19636(D^4 - d^4)/D$$

SHAFT FORMULAS — *continued*

solid shaft,

$$Z_P = 0.19636D^3$$

ratio of diameters of hollow shaft:

$$c = d/D$$

safety factor (sf): a number considered to be safe for the design, or as specified by an agency code for public safety, in that application.

shear strength, ultimate (S_U), for the material: STRENGTH OF MATERIALS

shear stress, working, (S_W):

$$S_W = S_U/\text{sf}$$
$$S_W = Tr/I_P = T/Z_P$$

hollow shaft,

$$S_W = 5.093TD/(D^4 - d^4)$$

solid shaft,

$$S_W = 5.093T/D^3$$

thickness of hollow shaft:

$$t = 0.5(D - d)$$

 torque:

$$T = S_W I_P/r = S_W Z_P$$
$$= 63{,}025(\text{hp})/N$$

hollow shaft,

$$T = 0.19636S_W(D^4 - d^4)/D$$

solid shaft,

$$T = 0.19636D^3S_W$$

torsional deflection: angle of twist   *
ultimate shear strength: shear strength, ultimate   *
working stress: shear stress, working, (S_W)   *

shear modulus: STRENGTH OF MATERIALS
shear stress: STRENGTH OF MATERIALS
sheet metal gauge thickness for brass and copper (in.):

40: 0.0031	25: 0.0179	10: 0.1019
39: 0.0035	24: 0.0201	9: 0.1144
38: 0.0040	23: 0.0226	8: 0.1285
37: 0.0045	22: 0.0253	7: 0.1443
36: 0.0050	21: 0.0285	6: 0.1620
35: 0.0056	20: 0.0320	5: 0.1819
34: 0.0063	19: 0.0359	4: 0.2043
33: 0.0071	18: 0.0403	3: 0.2294
32: 0.0080	17: 0.0453	2: 0.2576
31: 0.0089	16: 0.0508	1: 0.2893
30: 0.0100	15: 0.0571	1/0: 0.3249
29: 0.0113	14: 0.0641	2/0: 0.3648
28: 0.0126	13: 0.0720	3/0: 0.4096
27: 0.0142	12: 0.0808	4/0: 0.4600
26: 0.0159	11: 0.0907	5/0: 0.5165

sheet metal gauge thickness for iron and steel (in.):

38: 0.0060	23: 0.0269	9: 0.1495
37: 0.0064	22: 0.0299	8: 0.1644
36: 0.0067	21: 0.0329	7: 0.1793
35: 0.0075	20: 0.0359	6: 0.1943
34: 0.0082	19: 0.0418	5: 0.2092
33: 0.0090	18: 0.0478	4: 0.2242
32: 0.0097	17: 0.0538	3: 0.2391
31: 0.0105	16: 0.0598	2: 0.2656
30: 0.0120	15: 0.0673	1: 0.2813
29: 0.0135	14: 0.0747	1/0: 0.3125
28: 0.0149	13: 0.0897	2/0: 0.3438
27: 0.0164	12: 0.1046	3/0: 0.3750
26: 0.0179	11: 0.1196	4/0: 0.4063
25: 0.0209	10: 0.1345	5/0: 0.4375
24: 0.0239		

sheet (paper) =

0.04 quire
0.002 ream

short ton: TON
Si: silicon (element)
sin; sine: TRIGONOMETRY
sine law: TRIGONOMETRY
sine wave: curve per equation,

$$y = \sin x$$

where

a = amplitude
b = wavelength
x = angle, rad

amplitude,

$$a = \sin (\pi/2)$$

wavelength,

$$b = 2\pi$$

single phase: ELECTRICAL
skein =

360 ft
109.728 m

slenderness ratio for columns: COLUMN LOAD LIMITS
sliding friction: FRICTION
sling type psychrometer: PSYCHROMETER
slug: g-pound, gee-pound; English gravitational system unit of mass
that when subjected to 1 lb (force) accelerates at 1 ft/s², where

g = gravitational acceleration, 32.174 ft/s^2
W = weight, lb

$$\text{slug} = W/g = 0.03108W$$

slug =

1 gee-pound	32.174 lb (mass)
14.5939 kg	

slug/cubic foot: English unit for absolute (dynamic) density.
slug/cubic foot =

0.5154 g/cm^3
515.4 kg/m^3

slug/foot/hour: English unit for absolute (dynamic) viscosity.
slug/foot/hour =

13.3 cP	13.3 MPa-s
0.1330 dyn-s/cm^2	0.0133 N-s/m^2
0.1330 g/cm/s	0.1330 P
0.0133 kg/m/s	0.0133 Pa-s
2.778 E$-$04 lb (force)-s/ft^2	0.0089 pdl-s/ft^2
0.0089 lb (mass)/ft/s	2.778 E$-$04 slug/ft/s

slug/foot/second: English unit for absolute (dynamic) viscosity.
slug/foot/second =

4.788 E$+$04 cP	32.174 lb (mass)/ft/s
478.8 dyn-s/cm^2	4.788 E$+$08 μP
478.8 g/cm/s	4.788 E$+$04 mPa-s
47.88 kg/m/s	47.88 N-s/m^2
4.88 kg-s/m^2	478.8 P
1 lb (force)-s/ft^2	47.88 Pa-s
0.0069 lb (force)-s/in.^{2}	32.174 pdl-s/ft^2

slug/square foot = 0.2234 lb/in.^{2}
slug-square foot: English unit for the moment of inertia of solids in terms of mass.
slug-square inch: English unit for the moment of inertia of solids in terms of mass.
Sm: samarium (element)

Sn: tin (element)
soda ash: sodium carbonate
sonic speed: VELOCITY OF SOUND
sound: unit of measure for sound is the decibel. It is the logarithmic ratio of the pressure energy of the sound (or noise) in the air to the reference pressure of threshold of hearing, where

  dB = decibels
  P_1 = referenced pressure, 1 E − 16 W/m^2
  P_2 = measured pressure, W/m^2

decibels using logarithms to base 10,

$$dB = 10 \log (P_2/P_1)$$

noise levels for various sounds (dB):

air hammer: 110	radio, TV (quiet): 40
automobile: 50	riveter: 130
automobile horn: 120	rocket launch: 180
conversation: 65	sandblasting: 110
elevated train: 100	threshold hearing: 0
factory: 70	thunder: 100
office: 65	turbo jet: 170
punch press: 110	traffic, heavy: 75
radio, TV (loud): 80	whisper: 20

specific gravity, gas: the ratio of the weight of a gas of a specific volume to the weight of an equal volume of dry air (at 32°F and standard atmospheric pressure), or the ratio of the molecular weight of the gas to the molecular weight of air.
specific gravity, liquid or solid: the ratio of the weight of a specific volume of the material to the weight of an equal volume of water (at 39.2°F and standard atmospheric pressure). In the metric system, the density or weight of water is 1 g/cm^3, therefore, specific gravity and density are the same.

specific gravity, of two or more combined materials: the combined effective specific gravity of a substance consisting of two or more materials is the total of the products in percent content and specific gravity of each component.

specific gravity of various gases and vapors (at 32°F and atmospheric pressure):

acetic acid: 2.072
acetone: 2.004
acetylene: 0.898
AIR: 1.00
alcohol: 1.601
ammonia: 0.592
argon: 1.377
benzene: 2.694
benzine: 2.50
benzol: 2.694
butane: 2.067
butylene: 1.94
camphor: 5.29
carbon dioxide: 1.53
carbon disulfide: 2.63
carbon monoxide: 0.967
carbon tetrachloride: 5.308
chlorine: 2.447
chloroform: 4.125
ethane: 1.038
ethanol: 1.590
ether: 2.565
ethyl alcohol: 1.590
ethyl chloride: 2.365
ethylene: 0.9674
fluorine: 1.33
formaldehyde: 1.04
formic acid: 1.588
freon F12: 4.173
    F22: 3.178

gasoline: 3.50
glycerine: 3.178
helium: 0.138
heptane: 3.458
hexane: 2.91
hydrochloric acid: 1.26
hydrocyanic acid: 0.933
hydrofluoric acid: 2.370
hydrogen: 0.0696
hydrogen chloride: 1.259
hydrogen fluoride: 0.691
hydrogen sulfide: 1.176
kerosene: 4.50
krypton: 2.889
manufacturer's gas: 0.45
mercury vapor: 6.940
methane: 0.554
methanol: 1.106
methyl alcohol: 1.106
methyl chloride: 1.785
methyl ether: 1.590
methyl ethyl ketone: 2.488
mineral spirits: 3.899
naphthalene: 4.423
naphtha VMP: 3.75
natural gas: 0.6717
neon: 0.696
nicotine: 5.60
nitric oxide: 1.037
nitrogen: 0.967

specific gravity of various gases and vapors — *continued*

nitrogen oxide: 1.039
nitroglycerine: 7.838
nitrous oxide: 1.519
octane: 3.942
oxygen: 1.105
phenol: 3.246
phosphorus: 4.285
producer's gas: 0.860
propane: 1.52
propylene: 1.452
steam (212°F): 0.6218

styrene: 3.594
sulfur dioxide: 2.211
toluene: 3.18
toluol: 3.10
trichloroethylene: 4.535
turpentine: 4.702
water vapor (212°F): 0.6218
wood alcohol: 1.106
xenon: 4.53
xylene: 3.660
xylol: 3.70

specific gravity of various materials:

acetaldehyde: 0.778
acetic acid: 1.05
acetone: 0.786
alabaster: 2.7
alcohol: 0.88
aluminum: 2.70
amber: 1.1
americium: 11.7
ammonia: 0.825
antimony: 6.63
arsenic: 5.73
asbestos: 2.46
ashes, cinder: 0.64
asphalt: 0.995
babbitt, tin type: 7.4
bakelite: 1.35
barium: 3.75
barytes, loose: 2.31
        solid: 4.25
bauxite: 2.55
beeswax: 0.96
benzene: 0.876

benzine: 0.74
benzol: 0.876
beryllium: 1.82
bismuth: 9.8
borax: 1.75
boron: 2.55
brass: 8.5
brick, common: 1.8
     fire: 2.2
     hard: 2.0
     paving: 2.4
     rubble: 2.25
brine (25% NaCl): 1.20
bromine: 3.13
bronze, aluminum: 7.7
       phosphor: 8.87
butyric acid: 0.96
cadmium: 8.65
calcium: 1.55
calcium carbonate: 2.82
calcium chloride: 2.15
camphor: 0.999

specific gravity of various materials — *continued*

camphor oil: 0.88
carbide: 0.80
carbolic acid: 1.07
carbon: 2.21
carbon disulfide: 1.26
carbon tetrachloride: 1.59
carborundum: 2.25
castor oil: 0.96
cement, portland, loose: 1.48
                packed: 1.76
                solid: 3.1
ceramics: 2.5
cerium: 6.9
cesium: 1.9
chalk: 2.5
charcoal, oak: 0.53
         pine: 0.40
chromium: 7.19
cinders, coal: 0.64
clay: damp: 1.76
    dry: 1.0
coal, anthracite: 1.52
    bituminous: 1.27
    lignite: 1.25
cobalt: 8.9
coconut oil: 0.90
coke: 1.35
columbium: 7.25
concrete: 2.41
constantan: 8.9
copper: 8.92
copper ore: 4.2
copper oxide: 3.05
cork: 0.26
cork board: 0.16
corn oil: 0.92

cotton flax: 1.49
cottonseed oil: 0.93
creosote: 1.07
diamond: 3.51
diesel oil: 0.90
dolomite: 2.9
duralumin: 2.79
earth: 1.54
emery: 4.0
erbium: 9.18
ethyl acetate: 0.89
ethyl alcohol: 0.79
ethyl bromide: 1.43
ethylene chloride: 1.28
ethylene glycol: 1.11
europium: 9.18
feldspar: 2.57
felt: 0.22
fiberboard: 0.24
fiberglass: 0.064
flaxseed: 0.72
flint: 2.6
flour, loose: 0.45
    packed: 0.75
fly ash: 1.2
formaldehyde: 1.075
formic acid: 1.2
freon F11: 1.48
    F12: 1.31
    F22: 1.20
fuel oil no. 1: 0.83
    no. 2: 0.845
    no. 3: 0.873
    no. 4: 0.90
    no. 5: 0.943
    no. 6: 0.987

specific gravity of various materials — *continued*

Fuller's earth: 0.48
gadolinium : 7.96
galena: 7.46
gallium: 5.92
gasoline: 0.70
germanium: 5.35
german silver: 8.43
glass: 2.50
glue (66% water): 1.1
glycerine: 1.26
gold: 19.3
grain: 0.77
grain alcohol: 0.79
granite: 2.64
graphite: 2.1
gravel: 1.76
gum, Arabic: 1.44
gypsum: 2.3
hafnium: 13.0
heptane: 0.681
hexane: 0.656
holmium: 10.1
hydraulic oil: 0.85
hydrochloric acid: 1.16
hydrocyanic acid: 0.69
hydrofluoric acid: 1.16
hydrogen chloride: 1.268
hydrogen peroxide: 1.46
ice: 0.913
inconel: 8.55
indium: 7.31
inerteen: 1.55
invar (64 Fe/36 Ni): 7.98
iodine: 4.93
iridium: 22.46

iron, cast: 7.22
   malleable: 7.32
   pure: 7.88
   wrought: 7.78
iron ore, hematite: 5.25
   limonite: 3.80
   magnetite: 5.05
iron slag: 2.74
isopropyl alcohol: 0.80
ivory: 1.85
jet fuel: 0.80
kerosene: 0.82
lanthanum: 6.16
lard oil: 0.915
lead: 11.4
lead ore, galena: 7.46
leather: 0.95
lignite: 1.25
limestone: 2.60
linseed oil: 0.93
lithium: 0.53
lutethium: 9.76
lye (33% water): 1.70
magnesia, 85%: 0.30
magnesium: 1.74
malt, dry: 0.56
manganese: 7.45
manganese ore, pyrolusite: 4.16
manganin: 8.52
marble: 2.72
mercury: 13.6
methanol: 0.788
methyl acetate: 0.93
methyl alcohol: 0.788
methyl chloride: 0.92

specific gravity of various materials — *continued*

methylene chloride: 1.26
methyl ethyl ketone: 0.83
mica: 2.84
micarta: 1.36
milk: 1.03
mineral oil: 0.91
mineral spirits: 0.80
mineral wool: 0.16
molasses: 1.4
molybdenum: 10.2
monel (67% Ni): 8.87
mortar: 1.5
muriatic acid: 1.20
naphtha: 0.85
naphtha VMP: 0.75
Neat's foot oil: 0.91
neodymium: 7.06
nichrome (80 Ni/20 Cr): 8.4
nickel: 8.9
nicotine: 1.01
niobium: 8.6
nitric acid: 1.50
nitroglycerine: 1.60
octane: 0.70
oil: 0.89
oil, crude: 0.90
    lubricating, heavy: 0.93
    lubricating, light: 0.85
oleic acid: 0.891
olive oil: 0.91
osmium: 22.6
paint: 1.44
palladium: 12
palm oil: 0.93
paper: 0.93

paraffin: 0.90
peanut oil: 0.92
peat: 0.80
pentane: 0.626
petroleum: 0.87
phenol: 1.07
phenolic, molded: 1.35
phosphor bronze: 8.8
phosphoric acid: 1.78
phosphorus, red: 2.3
                yellow: 1.83
piano wire: 7.83
pine oil: 0.86
pitch: 1.12
plaster: 1.7
plaster of Paris: 2.2
platinum: 21.4
plexiglas: 1.19
plutonium: 19.7
polonium: 9.26
polyethylene: 0.915
polystyrene, molded: 1.07
porcelain: 2.4
potash: 0.96
potassium: 0.86
potatoes: 0.67
praseodymium: 6.64
propane: 0.509
propyl alchol: 0.80
propylene: 0.513
Prussic acid: 0.69
pumice stone: 0.90
pyralin: 1.35
pyrite, copper ore: 4.2
quartz: 2.6

specific gravity of various materials — *continued*

quenching oil: 0.91
radium: 5.0
resin: 1.08
rhenium: 20.5
rhodium: 12.44
rock salt: 2.17
rockwool: 0.27
rubber, hard: 1.22
    soft: 0.92
rubidium: 1.53
ruthenium: 12.2
salicylic acid: 1.44
salt: 2.1
sand: 1.9
sandstone: 2.3
sawdust: 0.40
scandium: 2.5
sea water: 1.03
selenium: 4.8
shale: 2.6
shellac: 1.2
silica: 2.89
silicon: 2.33
silver: 10.5
slag, crushed: 1.28
    furnace: 0.96
    ground: 0.37
slate: 2.8
snow, loose: 0.13
    wet: 0.80
soap powder: 0.36
soapstone: 2.7
soda ash: 0.48
sodium: 0.97
sodium carbonate: 2.43

sodium choride: 2.17
sodium hydroxide: 1.02
sodium nitrate: 2.25
sodium silicate: 1.44
solder (50 Pb/50 Sn): 9.3
soybean oil: 0.924
starch: 1.52
stearic acid: 0.847
steel, cast: 7.81
    silicon: 7.65
    stainless: 7.92
Stoddard solvent: 0.80
stone, crushed: 1.6
    solid: 2.4
strontium: 2.6
styrene: 0.91
sugar: 1.68
sulfur: 2.0
sulfuric acid: 1.83
talc: 2.7
tallow: 0.91
tantalum: 16.6
tar: 1.00
tartaric acid: 1.67
technetium: 11.5
tellurium: 6.24
terbium: 8.34
thallium: 11.87
thorium: 11.4
thulium: 9.37
tile: 1.2
tin: 7.3
tin ore, cassiterite: 6.7
titanium: 4.49
toluene: 0.86

specific gravity of various materials — *continued*

toluol: 0.898
trichloroethylene: 1.46
tung oil: 0.94
tungsten: 19.3
turpentine: 0.87
uranium: 18.7
urea: 1.50
vanadium: 5.9
varnish: 0.83
vegetable oil: 0.93
vinegar: 1.08
WATER: 1.000
whale oil: 0.92
WOOD: aborvitae: 0.29
     ash: 0.67
     balsam: 0.40
     bamboo: 0.38
     basswood: 0.35
     beech: 0.75
     birch: 0.64
     butternut: 0.42
     cedar: 0.37
     cherry: 0.64
     chestnut: 0.59
     cottonwood: 0.45
     cypress: 0.48
     ebony: 1.23
     elm: 0.58
     fir, Douglas: 0.51
     gum: 0.55
     hemlock: 0.45
     hickory: 0.80

WOOD — *continued*

locust: 0.72
magnolia: 0.54
mahogany: 0.56
maple, red: 0.71
     sugar: 0.61
     white: 0.53
oak, red: 0.67
     white: 0.77
pine, white: 0.43
     yellow: 0.54
poplar: 0.48
redwood: 0.42
spruce: 0.40
sycamore: 0.56
tamarack: 0.50
tupelo: 0.51
walnut: 0.61
willow: 0.48
wood alcohol: 0.79
wool: 1.31
xylene: 0.90
xylol: 0.87
ytterbium: 7.0
yttrium: 5.52
zinc: 7.14
zinc chloride: 2.9
zinc ore, blende: 4.0
zinc oxide: 5.6
zircon: 4.7
zirconium: 6.5

specific heat: HEAT TRANSFER
specific humidity: HUMIDITY, SPECIFIC
specific viscosity: FLUID FLOW
specific volume: VOLUME, SPECIFIC

specific weight: WEIGHT, SPECIFIC

specific weight of various gases and vapors: WEIGHTS (SPECIFIC) OF VARIOUS GASES AND VAPORS

specific weight of various liquids: WEIGHTS (SPECIFIC) OF VARIOUS LIQUIDS

specific weight of various materials: WEIGHTS (SPECIFIC) OF VARIOUS MATERIALS

speed of light: VELOCITY OF LIGHT

speed of sound, sonic speed: VELOCITY OF SOUND

sphere = 12.566 sr

SPHERE FORMULAS: where

A = area, ft^2
c = chord length, ft
cg = center of gravity
d = inside diameter, ft
D = outside diameter, ft
h = height of spherical segment, ft
I_{AA} = moment of inertia, slug-ft^2
k_{AA} = radius of gyration, ft
r = radius, ft
V = volume, ft^3
W = weight, lb

sphere, hollow:

area of surface,

$$A = 3.1416D^2$$

moment of inertia,

$$I_{AA} = 0.00311W(D^5 - d^5)/(D^3 - d^3)$$

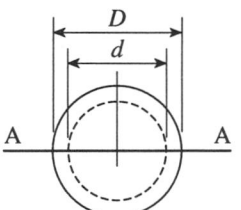

radius of gyration,

$$k_{AA} = 0.3162(D^5 - d^5)^{1/2}/(D^3 - d^3)^{1/2}$$

volume,

$$V = 0.5236(D^3 - d^3)$$

SPHERE FORMULAS — *continued*

sphere, solid:

area of surface,

$$A = 3.1416D^2$$

moment of inertia,

$$I_{AA} = 0.00311D^2W$$

radius of gyration,

$$k_{AA} = 0.3162D$$

volume,

$$V = 0.5236D^3$$

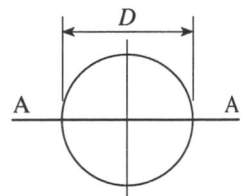

sphere, thin hollow:

area of surface,

$$A = 3.1416D^2$$

moment of inertia,

$$I_{AA} = 0.00518D^2W$$

radius of gyration,

$$k_{AA} = 0.408D$$

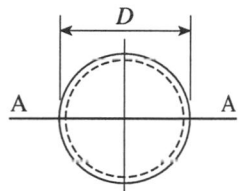

spherical sector, solid:

area of surface,

$$A = 1.571r(c + 4h)$$

center of gravity (cg),

$$y = 0.375(2r - h)$$

chord length,

$$c = 2(2rh - h^2)^{1/2}$$

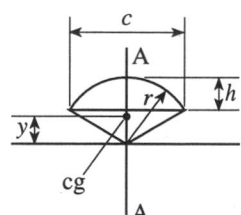

SPHERE FORMULAS — *continued*

spherical sector, solid — *continued*

height of spherical segment,

$$h = r - 0.5(4r^2 - c^2)$$

moment of inertia,

$$I_{AA} = 0.0062Wh(3r - h)$$

radius of gyration,

$$k_{AA} = 0.775(3rh - h^2)^{1/2}$$

volume,

$$V = 2.0944r^2h$$

spherical segment, solid:

area of surface,

$$A = 6.283rh$$

center of gravity (cg),

$$y = 0.75(2r - h)^2/(3r - h)$$

chord length,

$$c = 2(2rh - h^2)^{1/2}$$

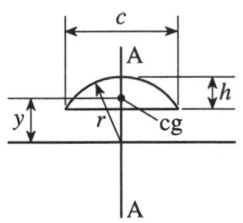

height of segment,

$$h = r - 0.5(4r^2 - c^2)$$

moment of inertia,

$$I_{AA} = 0.0622Wh(r^2 - 0.75rh + 0.15h^2)/(3r - h)$$

radius of gyration,

$$k_{AA} = 5.67(I_{AA}/W)^{1/2}$$

volume,

$$V = 1.0472h^2(3r - h)$$

SPRING FORMULAS: helical springs formed from round wire into a continuous series of coil, where

c = spring index
d = wire diameter, in.
D = mean diameter of coil, in.
D_O = outside diameter of coil, in.
f = deflection, in.
f_R = deflection rate, lb/in.
F = load, lb
G = modulus of elasticity, torsion, lb/in.^{2}
k = Wahl factor
N = number of active coils
s = shear stress, maximum, lb/in.^{2}

coil mean diameter:

$$D = D_O - d$$

deflection, total:

$$f = 8D^3FN/d^4G$$

deflection rate:

$$f_R - 0.125d^4G/D^3N$$

load capacity of spring:

$$F = 0.125d^4fG/D^3N$$

modulus of elasticity, torsional:

$$G = 8D^3FN/d^4f$$

number of active coils:

$$N = 0.125d^4fG/D^3F$$

shear stress, maximum:

$$s = 2.55DF/d^3$$

with Wahl factor:

$$s = 2.55DFk/d^3$$

SPRING FORMULAS—*continued*

spring index:

$$c = D/d$$

Wahl factor: stress concentration factor to predict the increase of stress as the spring index decreases,

$$k = [(4c - 1)/(4c - 4)] + (0.615/c)$$

wire diameter:

$$d = 1.366(DFk/s)^{1/3}$$

---

spur gear: GEAR FORMULAS

square: four-sided figure with all equal sides.

area,

$$A = a^2$$

area of square (sides a) equal to area of circle (diameter D),

$$a = 0.88623D$$

diagonal,

$$d = 1.414a$$

diameter of circle enclosing a square,

$$d = 1.414a$$

square inscribed within a circle with diameter d,

$$a = 0.7071d$$

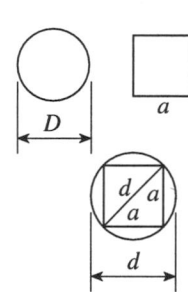

SQUARE, SECTION FORMULAS: where

$$a, x, y, z = \text{dimensions, in.}$$
$$A = \text{area, in.}^2$$
$$cg = \text{center of gravity}$$
$$I_{AA}, I_{BB}, I_{CC}, I_{DD} = \text{moment of inertia, in.}^4$$
$$I_P = \text{polar moment of inertia, in.}^4$$
$$k_{AA}, k_{BB}, k_{CC}, k_{DD} = \text{radius of gyration, in.}$$

SQUARE, SECTION FORMULAS—*continued*

$$k_P = \text{polar radius of gyration, in.}$$
$$s = \text{perimeter, in.}$$
$$Z_{AA}, Z_{BB}, Z_{CC}, Z_{DD} = \text{section modulus, in.}^3$$
$$Z_P = \text{polar section modulus for torsion, in.}^3$$

square, hollow:

area,

$$A = a^2 - b^2$$

center of gravity (cg),

$$x = y = 0.5a$$
$$z = 0.707a$$

moment of inertia,

$$I_{AA} = I_{BB} = I_{CC} = 0.0833(a^4 - b^4)$$
$$I_{DD} = 0.0833[4a^4 - b^2(3a^2 + b^2)]$$

moment of inertia, polar,

$$I_P = 0.1667(a^4 - b^4)$$

radius of gyration,

$$k_{AA} = k_{BB} = k_{CC} = 0.2887(a^2 + b^2)^{1/2}$$
$$k_{DD} = 0.577(a^2 - b^2)^{1/2}$$

section modulus,

$$Z_{AA} = Z_{BB} = 0.1667(a^4 - b^4)/a$$
$$Z_{CC} = 0.118(a^4 - b^4)/a$$
$$Z_{DD} = 0.333(a^4 - b^4)/a$$

square, solid:

area,

$$A = a^2$$

center of gravity (cg),

$$x = y = 0.5a$$
$$z = 0.707a$$

SQUARE, SECTION FORMULAS — *continued*

moment of inertia,

$$I_{AA} = I_{BB} = I_{CC} = 0.0833a^4$$
$$I_{DD} = 0.3333a^4$$

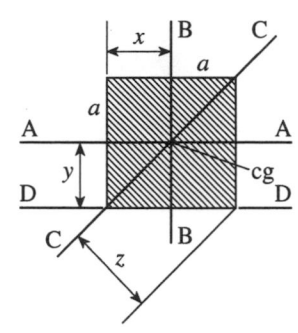

moment of inertia, polar,

$$I_P = 0.1667a^4$$

perimeter,

$$s = 4a$$

radius of gyration,

$$k_{AA} = k_{BB} = k_{CC} = 0.2887a$$
$$k_{DD} = 0.5774a$$

radius of gyration, polar,

$$k_P = 0.4082a$$

section modulus,

$$Z_{AA} = Z_{BB} = 0.1667a^3$$
$$Z_{CC} = 0.118a^3$$
$$Z_{DD} = 0.3333a^3$$
$$Z_P = 0.222a^3$$

---

square centimeter =

2.471 E − 08 acres	0.1550 in.^{2}
1 E − 06 ares	1 E − 10 km^2
1 E − 04 centares	1 E − 04 m^2
0.1974 circular in.	3.861 E − 11 mi^2
127.324 circular mm	1.55 E + 05 mil^2
1.97352 E + 05 cmil	100 mm^2
0.01 dm^2	3.954 E − 06 rod^2
0.001076 ft^2	1.196 E − 04 yd^2
1 E − 08 ha	

square centimeter/dyne/second: RHE

square centimeter/second: metric unit of kinematic viscosity.
square centimeter/second: STOKE
square decimeter =

100 cm^2	15.5 in.^{2}
0.1076 ft^2	0.01 m^2

square dekameter =

0.02471 acre	1.55 E + 05 in.^{2}
1 are	100 m^2
100 centares	3.861 E − 05 mi^2
1 E + 06 cm^2	3.954 rods^2
0.001076 ft^2	119.6 yd^2
0.01 ha	

square foot, square feet =

2.296 E − 05 acre	144 in.^{2}
9.29 E − 04 are	9.29 E − 08 km^2
0.0929 centare	0.0929 m^2
929.0 cm^2	3.587 E − 08 mi^2
1.833 E + 08 cmil	9.29 E + 04 mm^2
9.2903 dm^2	0.003673 rod^2
9.29 E − 06 ha	0.1111 yd^2

square foot/hour: English unit of kinematic viscosity.
square foot/hour =

25.8 cSt	0.04 in.^{2}/s
0.258 cm^2/s	2.58 E − 05 m^2/s
2.778 E − 04 ft^2/s	25.8 mm^2/s
144 in.^{2}/h	0.258 St

square foot/second: English unit of kinematic viscosity.
square foot/second =

9.29 E + 04 cSt	144 in.^{2}/s
929 cm^2/s	0.0929 m^2/s
3600 ft^2/h	9.29 E + 04 mm^2/s
5.18 E + 05 in.^{2}/h	929.37 St

square hectometer = 1 E + 04 m²
square inch =

1.594 E − 07 acre	6.452 E − 08 ha
6.452 E − 06 are	6.452 E − 10 km²
6.452 E − 04 centare	6.452 E − 04 m²
1.27324 circular in.	2.491 E − 10 mi²
821.5 circular mm	1 E + 06 mil²
6.452 cm²	645.2 mm²
1.27324 E + 06 cmil	2.55 E − 05 rod²
0.064516 dm²	7.716 E − 04 yd²
0.00694 ft²	

square inch/hour: English unit of kinematic viscosity.
square inch/hour =

0.00179 cm²/s	2.778 E − 04 in.²/s
0.1792 cSt	1.792 E − 07 m²/s
0.0069 ft²/h	0.1792 mm²/s
1.929 E − 06 ft²/s	0.001792 St

square inch/second: English unit of kinematic viscosity.
square inch/second =

2.3226 E + 04 cm²/h	0.006944 ft²/s
6.45 cm²/s	3600 in.²/h
645.16 cSt	6.452 E − 04 m²/s
25 ft²/h	645 mm²/s
0.4167 ft²/min	6.4516 St

square kilometer =

247 acres	1.55 E + 09 in.²
1 E + 10 cm²	1 E + 06 m²
1.0764 E + 07 ft²	0.3861 mi²
100 ha	1.196 E + 06 yd²

square meter =

2.471 E−04 acre	1 E−04 hm^2
0.01 are	1550 in.^{2}
1 centare	1 E−06 km^2
1 E+04 cm^2	3.861 E−07 mi^2
0.01 dam^2	1 E+06 mm^2
100 dm^2	0.03954 rod^2
10.764 ft^2	1.196 yd^2
1 E−04 ha	

square meter/hour: metric unit of kinematic viscosity.
square meter/hour =

2.7778 cm^2/s	0.43056 in.^{2}/s
0.00299 ft^2/s	2.7778 E−04 m^2/s

square meter/newton-second = 0.1 rhe
square meter/second: metric unit of kinematic viscosity.
square meter/second =

1 E+04 cm^2/s	5.58 E+06 in.^{2}/h
1 E+06 cSt	1550 in.^{2}/s
3.88 E+04 ft^2/h	1 E+06 mm^2/s
10.764 ft^2/s	1 E+04 St

square mil =

6.452 E−06 cm^2	1 E−06 in.^{2}
1.2732 cmil	6.452 E−04 mm^2

square mile =

640 acres	4.0145 E+09 in.^{2}
25,900 ares	2.590 km^2
2.59 E+10 cm^2	2.590 E+06 m^2
2.78783 E+07 ft^2	1.024 E+05 rods^2
259 ha	3.098 E+06 yd^2

square millimeter =

1.27324 circular mm	0.00155 in.^{2}
0.01 cm^2	1 E − 06 m^2
1973.52 cmil	1550 mil^2
1.076 E − 05 ft^2	1.196 E − 06 yd^2

square millimeter/second: metric unit of kinematic viscosity.
square millimeter/second: CENTISTOKE
square perch: SQUARE ROD
square rod =

0.00625 acre	39,204 in.^{2}
0.253 are	25.293 m^2
2.5293 E + 05 cm^2	9.766 E − 06 mi^2
272.25 ft^2	1 perch^2
0.00253 ha	30.25 yd^2

square yard =

2.066 E − 04 acre	1296 in.^{2}
0.008361 are	8.36 E − 07 km^2
0.83613 centare	0.8361 m^2
8361 cm^2	3.228 E − 07 mi^2
0.00836 dam^2	8.361 E + 05 mm^2
9 ft^2	0.03306 rod^2
8.36 E − 05 ha	

Sr: strontium (element)
SSF: Saybolt Second Furol
SSU: Saybolt Second Universal, SUS
stainless steel 18-8: stainless steel containing 18% chromium and 8% nickel.
standard conditions (air or gas): AIR FLOW; GAS FLOW
standard cubic feet per minute (scfm): AIR FLOW; GAS FLOW
statampere: metric unit for current as used in the electrostatic system of units.

statampere $=$

    3.336 E $-$ 10 A   3.336 E $-$ 11 abampere

statampere/square centimeter $= 2.152$ E $- 09$ A/in.^{2}
statcoulomb: ELECTROSTATICS
statcoulomb $-$

    3.336 E $-$ 11 abcoulombs   2.0822 E $+ 09$ electronic charges
    9.266 E $-$ 14 A-h   3.4573 E $-$ 15 faraday
    3.336 E $-$ 10 C

statcoulomb/square centimeter $= 2.15$ E $- 09$ C/in.^{2}
statfarad: metric unit for capacitance as used in the electrostatic system
    of units.
statfarad $=$

    1.113 E $-$ 21 abfarad   1.113 E $- 06$ μF
    1.113 E $-$ 12 F

stathenry: metric unit for inductance as used in the electromagnetic
    system of units.
stathenry $-$

    8.988 E $+ 20$ abhenrys   8.988 E $+ 17$ μH
    8.988 E $+ 11$ H   8.988 E $+ 14$ mH

static electricity: electrical charges produced in a body by friction and
    that remain at rest in the body until removed by contact with
    another body.
static friction coefficient: FRICTION
static head: AIR FLOW; FLUID FLOW; GAS FLOW.
static pressure head: AIR FLOW; FLUID FLOW; GAS FLOW.
STATICS: study of forces acting on a body not in motion, where

                 a, b = angle of force, degrees
                     C = couple effect, ft-lb
                     d = distance, ft

STATICS — *continued*

F = force, lb
F_H = horizontal component of force, lb
F_R = resultant force, lb
F_V = vertical component of force, lb
F_1, F_2 = individual forces, lb
M = moment, ft-lb
s = distance between forces, ft

couple: two equal forces acting on a body in opposite directions and located a specific distance apart, producing a turning effect on the body,

$$C = Fs$$

equilibrium: when summation of forces in all directions is zero, at a specific point.

force, resultant: direction and magnitude of a single force that is equivalent to several forces in the same plane acting on the same point.

forces in the same direction,

$$F_R = F_1 + F_2 + F_3$$

forces in opposite directions,

$$F_R = F_1 - F_2$$

two forces 90° apart,

$$F_R = (F_1^2 + F_2^2)^{1/2}$$

two forces $a°$ apart,

$$F_R = (F_1^2 + F_2^2 + 2F_1F_2 \cos a)^{1/2}$$
$$\sin b = (F_1 \sin a)/F_R$$

force components:

$$F_H = F \cos a$$
$$F_V = F \sin a$$

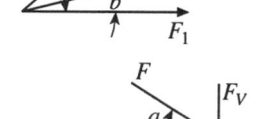

STATICS—*continued*

moment: torque turning effect of a force about a point where the moment equals the product of the force times the perpendicular distance from the point.

$$M = Fd$$

reaction: the opposing force to an applied force on a body. In general, the sum of reactions of a body is equal to all forces acting on the body.

statmho =

1.113 E−21 abmho	1.113 E−12 mho
1.113 E−18 megmho	1.113 E−06 micromho

statmho/centimeter; statmho/centimeter cube: metric unit of electrical conductivity of a specific material with cross-sectional area of 1 cm^2 in statmhos per centimeter length.

statmho/centimeter = 1.113 E−12 mho/cm

statmho/centimeter cube: STATMHO/CM

statoersted = 3.336 E−11 Oe

statohm: metric unit for resistance as used in the electrostatic system of units.

statohm =

8.988 E+20 abohms	8.988 E+17 microhms
8.988 E+05 megohms	8.988 E+11 ohms

statohm/centimeter cube: STATOHM-CENTIMETER

statohm-centimeter; statohm-centimeter cube; statohm (centimeter cube); statohm/centimeter cube: metric unit of electrical resistivity of a specific material with cross-sectional area of 1 cm^2 in statohms per unit centimeter length.

statohm-centimeter = 8.988 E+11 ohm-cm

statohm-centimeter cube: STATOHM-CENTIMETER

statohm (centimeter cube): STATOHM-CENTIMETER
statute mile: MILE
statvolt: metric unit for potential as used in the electrostatic system of units.
statvolt =

2.998 E + 10 abvolts	299.8 V
2.998 E + 05 mV	

statvolt/centimeter =

2.998 E + 10 abvolts/cm	299.8 V/cm
0.2998 kV/cm	761.47 V/in.
2.998 E + 05 mV/cm	0.761 V/mil

statvolt/inch = 118 V/cm
statweber = 299.8 Wb
statweber/square centimeter = 2.998 E + 10 G
STEAM FLOW: where

a = coefficient of thermal linear expansion, length change/unit length/°F
A = area of pipe, in.^{2}
C_F = coefficient of pipe fitting
D = inside diameter of pipe, in.
E = internal energy of steam, ft-lb/lb
E_K = kinetic energy, ft-lb/lb
E_P = potential energy, ft-lb/lb
E_W = work energy, ft-lb/lb
f = friction factor
F = flow rate by volume, ft^3/min
H = enthalpy, Btu/lb
L = length of pipe, ft
L_C = thermal expansion of pipe length, ft
L_E = pipe length equivalent, ft
N_R = Reynold's number
p = gauge pressure, psig
$p_1 - p_2$ = pressure drop, psig
P = absolute pressure, psia
P_1 = initial pressure, psia

STEAM FLOW — *continued*

P_2 = final pressure, psia
P_L = pressure loss, psia
P_T = critical pressure, psia
Q = heat, Btu/lb
R = gas constant, ft-lb/lb/°R
t = temperature, °F
T = absolute temperature, °R
v = specific volume, ft³/lb
V = velocity of steam, ft/min
w = specific weight of steam, lb/ft³
W = flow rate by weight, lb/min
μ = absolute viscosity, cP
v = kinematic viscosity, cSt

absolute pressure, psia:

$$P = p + 14.696$$

absolute temperature, °R:

$$T = t + 459.69$$

absolute viscosity: VISCOSITY, ABSOLUTE
absolute viscosity, steam: $\mu = 0.011$ at 212°F
area cross-section of pipe:

$$A = 0.7854D^2$$

coefficient of pipe fitttings (C_F):

for conversion to length equivalent (L_E):

elbow, 90°: 0.8	valves, open, angle: 4.0
45°: 0.55	check: 2.3
sweep: 0.63	gate: 0.17
tee: 1.63	globe: 8.3

coefficient of viscosity: VISCOSITY, ABSOLUTE
coefficient of viscosity, steam: absolute viscosity, steam *
conservation of energy in steady flow system between points 1 and 2:

$$E_1 + E_{K1} + E_{P1} + 778Q = E_2 + E_{K2} + E_{P2} + E_W$$

STEAM FLOW — *continued*

critical pressure (P_T): pressure at maximum flow rate.

for dry saturated steam,

$$P_T = 0.58P_1$$

for superheated steam,

$$p_T = 0.55p_1$$

if $P_2 < P_T$, discharge rate is constant for an unretarded flow.
if $P_2 > P_T$, discharge rate decreases as P_2 increases for a retarded flow.

dynamic viscosity: VISCOSITY, ABSOLUTE
enthalpy: internal energy of the steam plus product of pressure-volume,

$$H = (E + 144Pv)/778$$

Enthalpy values (Btu/lb) for specific pressure and temperature conditions are found in Steam Tables or on a Mollier diagram.
flow rate, laminar flow:

by volume,

$$F = 0.0069AV$$
$$= 0.00545D^2V$$

by weight,

$$W = 0.0069AVw$$
$$= 0.00545D^2Vw$$

flow rate, turbulent flow:

by volume,

$$F = 19.3[R(P_1^2 - P_2^2)D^5/fTL]^{1/2}$$

by weight,

$$W = 77[(P_1^2 - P_2^2)D^5/fRTL]^{1/2}$$
$$= 8.323[(P_1^2 - P_2^2)D^5/fTL]^{1/2}$$

STEAM FLOW — *continued*

flow rate types, based on Reynold's number:

laminar flow, $N_R < 2000$
transitional flow, $2000 < N_R < 4000$
turbulent flow, $N_R > 4000$

friction factor: may be obtained from a Moody diagram or other charts.

friction factor for laminar flow,

$$f = 64/N_R$$

gas constant (R) for steam: $R = 85.58$ ft-lb/lb/°R
heat of vaporization: approximately 970 Btu of heat per pound are required to convert water to steam.
internal energy (E): stored energy as related to the temperature and pressure of the steam. It cannot be measured; however, the change in internal energy may be measured between points 1 and 2, as $E_1 - E_2$.
kinetic energy: energy due to the velocity of the steam,

$$E_K = 4.317 \text{ E} - 06 \ V^2$$

kinematic viscosity: VISCOSITY, KINEMATIC
laminar flow: LAMINAR FLOW
length equivalent (L_E) of pipe fittings:

$$L_E = 2.75 C_F D$$

Moody diagram: MOODY DIAGRAM
potential energy (E_P): energy due to the elevation of the system above datum.
pressure drop due to friction:

laminar flow,

$$p_1 - p_2 = 1.114 \text{ E} - 05 \ \mu(L + L_E)V/D^2$$

turbulent flow,

$$p_1 - p_2 = 3.6 \text{ E} - 07 f(L + L_E)V^2 w/D$$

STEAM FLOW — *continued*

 Reynold's number:

$$N_R = 2.07VDw/\mu = 129VD/v$$

specific volume, as found in Steam Tables for specific pressure and temperature:

$$v = 1/w$$

specific weight, as found in Steam Tables for specific pressure and temperature:

$$w = 1/v$$

thermal expansion of pipe length between points 1 and 2: (*Note:* $a = 6.5$ E−06 for steel pipe.)

$$L_C = 12aL(t_2 - t_1)$$

transitional flow: TRANSITIONAL FLOW
turbulent flow: TURBULENT FLOW
velocity of flow between points 1 and 2:

$$V = 13{,}428(H_1 - H_2)^{1/2}$$
$$= 183.35W/D^2w$$

weight, specific: specific weight   *
work energy at constant pressure ($P_1 = P_2$):

$$E_W = 144P(v_2 - v_1)$$

Steel Wire Gauges (American Wire Gauge)

Size	Dia. (in.)	Weight (lb/1000-ft)	Size	Dia. (in.)	Weight (lb/1000-ft)
50	0.0044	0.0512	21	0.0318	2.67
49	0.0046	0.0560	20	0.0348	3.20
48	0.0048	0.0609	19	0.0410	4.45
47	0.0050	0.0661	18	0.0475	5.96
46	0.0052	0.0715	17	0.0540	7.71
45	0.0055	0.0800	16	0.0625	10.33
44	0.0058	0.0890	15	0.0720	13.7
43	0.0060	0.0952	14	0.0800	16.9
42	0.0062	0.1017	13	0.0915	22.1
41	0.0066	0.1152	12	0.1055	29.4
40	0.0070	0.1296	11	0.1205	38.4
39	0.0075	0.1488	10	0.1350	48.2
38	0.0080	0.1693	9	0.1483	58.2
37	0.0085	0.1911	8	0.1620	69.4
36	0.0090	0.214	7	0.1770	82.9
35	0.0095	0.239	6	0.1920	97.5
34	0.0104	0.286	5	0.2070	113.3
33	0.0118	0.368	4	0.2253	134.2
32	0.0128	0.433	3	0.2437	157.1
31	0.0132	0.461	2	0.2625	182.3
30	0.0140	0.518	1	0.2830	211.8
29	0.0150	0.595	1/0	0.3065	248.5
28	0.0162	0.694	2/0	0.3310	289.8
27	0.0173	0.792	3/0	0.3625	347.6
26	0.0181	0.866	4/0	0.3938	410.2
25	0.0204	1.10	5/0	0.4305	490.2
24	0.0230	1.40	6/0	0.4615	563.3
23	0.0258	1.76	7/0	0.4900	635.1
22	0.0286	2.16			

Stefan–Boltzmann's constant: HEAT TRANSFER
Stefan–Boltzmann's law: HEAT TRANSFER
steradian: ILLUMINATION
steradian =

0.1592 hemisphere
0.07958 sphere

stere: CUBIC METER
stoke: cm^2/s; metric unit of kinematic viscosity; equals absolute viscosity/density.

stoke =

100 cSt	0.155 in.^{2}/s
1 cm^2/s	1 E$-$04 m^2/s
3.88 ft^2/h	100 mm^2/s
0.001076 ft^2/s	0.00226SSU $-$ 1.95/SSU, SSU < 100 s
558 in.^{2}/h	0.00220SSU $-$ 1.35/SSU, SSU > 100 s

strain: STRENGTH OF MATERIALS
STRENGTH OF MATERIALS: where

a = angle of twist, rad
A = area, in.^{2}
c = distance from neutral axis to outer fiber of the beam section, in.
e = strain, in./in.
E = modulus of elasticity, compression or tension, lb/in.^{2}
F = force (load), lb
G = modulus of elasticity, shear, lb/in.^{2}
I = moment of inertia, in.^{4}
I_P = polar moment of inertia, in.^{4}
l = length change, in.
L = length, in.
M = maximum moment, in.-lb
R = reaction, lb
S = stress, lb/in.^{2}
S_S = shear stress, lb/in.^{2}
S_U = ultimate strength of a material, lb/in.^{2}

STRENGTH OF MATERIALS — *continued*

S_W = working stress, lb/in.^{2}
sf = safety factor, code or design
T = torque, in.-lb
Z = section modulus, in.^{3}
Z_P = polar section modulus, in.^{3}

allowable stress: stress, working   *

axial load: tension or compression force applied in line with the length of the body.

beam formulas: BEAM LOAD FORMULAS

bending strength, maximum, of various materials: ultimate strength (bending) of various materials   *

bending stress: stress, bending   *

change in length: elongation   *

column loading formulas: COLUMN LOAD LIMITS; COLUMN LOADING FORMULAS

compression strength, maximum, of various materials: ultimate strength (compression) of various materials   *

compressive stress: stress, compressive   *

deflection: measure of amount of bending in a beam as a result of load applied to the beam.

deflection formulas, beams: BEAM LOAD FORMULAS

elasticity: property of a material that allows it to return to its original shape after the load is removed.

elastic limit; proportional limit: maximum stress that a material may be distorted under a load and return to its original shape after the load is released. The point where stress/strain ratio changes from a constant value and strain (elongation) increases at a higher rate than stress (loading).

elastic limit (bending) of various wood materials (1000 lb/in.^{2}):

cedar, red: 5.5	pine, white: 6.3
cypress: 6.5	yellow: 5.0
fir: 6.5	redwood: 4.5
hemlock: 6.3	spruce: 8.0
oak, white: 3.5	tamarack: 7.5

STRENGTH OF MATERIALS — *continued*

elastic limit (compression) of various materials (1000 lb/in.^{2}):

aluminum, cast: 9
    drawn: 20
brass, rolled: 25
bronze, cast: 20
concrete: 1
copper, cast: 8
    drawn: 38

iron, cast: 20
    malleable: 25
steel, cold-rolled: 60
    h.r. high-carbon:70
    h.r. low-carbon: 35
    structural: 36

elastic limit (shear) of various materials (1000 lb/in.^{2}):

brass, rolled: 15
copper, drawn: 23
iron, malleable: 23
    wrought: 8

steel, cold-rolled: 36
    h.r. high-carbon: 42
    h.r. low-carbon: 21

elastic limit (tension) of various materials (1000 lb/in.^{2}):

aluminum, cast: 6.5
    drawn: 20
    rolled: 13
aluminum/bronze: 40
alum wire, annealed: 14
brass, cast: 6
    rolled: 25
brass wire, annealed: 16
bronze, cast: 20
    rolled: 40
bronze/phosphor: 24
copper, cast: 6
    drawn: 38
    rolled: 10
copper wire, annealed: 10
german silver: 19

gold, cast: 4
iron, annealed: 27
    cast: 6
    malleable: 36
    wrought: 26
magnesium, extruded: 17
stainless steel: 35
steel, cast: 30
    cold-rolled: 60
    h.r. high-carbon: 70
    h.r. low-carbon: 35
    structural: 33
steel wire: 60
steel wire, annealed: 40
tin, cast: 1.6
zinc, cast: 4

elongation: change in length of a body due to axial (tensile or compression) load applied,

$$l = FL/AE$$

STRENGTH OF MATERIALS — *continued*

fatigue failure: failure (rupture) of a material when subjected to frequent loading and repeated stressing.

force:

$$F = AS$$

Hooke's law: stress is directly proportional to strain in a material, according to its modulus of elasticity,

$$E = S/e$$

length change: elongation   *

modulus of elasticity (E): Young's modulus; ratio of stress to strain of a material in compression or tension loading, up to the proportional limit of the material,

$$E = S/e$$
$$= FL/Al$$

modulus of elasticity (bending) of various wood materials (1 E + 06 lb/in.^{2}):

cedar, red: 1.0	pine, white: 1.0
cypress: 1.2	yellow: 4.5
fir: 1.6	redwood: 1.2
hemlock: 1.4	spruce: 1.2
oak, white: 1.7	tamarack: 1.3

modulus of elasticity (compression) of various materials (1E + 06 lb/in.^{2}):

aluminum: 10	iron, cast: 20
brass: 13	limestone: 8
brick: 2	magnesium, extruded: 6.5
bronze, cast: 12	marble: 8
concrete: 3.3	sandstone: 3
copper: 16	slate: 14
granite: 7	steel: 30

STRENGTH OF MATERIALS — *continued*

modulus of elasticity (shear) of various materials (1 E + 06 lb/in.^{2}):

aluminum: 4
brass: 5
bronze: 5
copper: 6
iron, cast: 6
   malleable: 12

iron, wrought: 10
lead: 0.8
magnesium, extruded: 2.4
stainless steel: 10
steel: 12

modulus of elasticity (tension) of various materials (1 E + 06 lb/in.^{2}):

aluminum: 10
brass: 13
bronze, cast, red: 12
    cast, yellow: 14
    rolled: 14
concrete: 2
copper: 16
gold, cast: 6
iron, cast: 5
   malleable: 25

iron, wrought: 27
lead, cast: 1.0
   rolled: 2.2
magnesium, extruded: 6.5
phenolic, molded: 0.9
platinum: 21.5
polystyrene: 5
stainless steel: 28
steel: 30
zinc: 13

modulus of elasticity, rigidity (*G*): modulus of elasticity, shear (*G*)   *
modulus of elasticity, shear (*G*): modulus of elasticity (torsion); modulus of rigidity; modulus of shear; shear modulus; stiffness; rigidity; torsion modulus: ratio of the shearing stress to the shearing strain of a material in torsion load,

$$G = S/e$$
$$= FL/Al$$
$$= TL/aI_P$$

modulus of elasticity, torsion (*G*): modulus of elasticity, shear (*G*)   *
modulus of rigidity: modulus of elasticity, shear (*G*)   *
modulus of rupture: ultimate strength   *
modulus of shear: modulus of elasticity, shear (*G*)   *
moment, maximum: algebraic summation of the moments of all loads (concentrated, distributed, rolling, etc.) acting on a beam, as used to determine bending stress and beam strength required for a given material,

$$M = S_U I/c = S_U/Z$$

STRENGTH OF MATERIALS — *continued*

moment of inertia: MOMENT OF INERTIA

neutral plane: plane through a beam where compression and tension stresses are zero.

Poisson's ratio: ratio of the relative change in the diameter of a round bar to its change of length under axial loading within the elastic limit of the material.

polar section modulus: section modulus, polar   *

proportional limit: elastic limit   *

radius of gyration: RADIUS OF GYRATION

reaction: opposition to a force or load, as in beam loading. Total reaction at the supports of a beam is equal to the total load or force on the beam,

$$R_1 + R_2 = F_1 + F_2$$

rigidity: modulus of elasticity (G) *

rupture: failure of a material, as from fatigue.

rupture point: ultimate strength   *

safety factor: a number considered to be safe for a design or as specified by an agency code for public safety in that application.

section modulus: ratio of the moment of inertia of the cross-sectional area of a beam about its neutral axis to the distance from the neutral axis to the extreme fiber,

$$Z = I/c$$

section modulus, polar:

$$Z_P = I_P/c$$

section modulus for various sections: refer to ANGLE; CHANNEL; CIRCLE; ELLIPSE; HEXAGON; OCTAGON; RECTANGLE; SQUARE; TRAPEZOID; TRIANGLE; Z BAR.

shear modulus: modulus of elasticity, shear (G) *

shear strength (ultimate) of various materials: ultimate strength (shear) of various materials   *

shear stress: stress, shear   *

stiffness: modulus of elasticity, shear (G) *

STRENGTH OF MATERIALS — *continued*

strain: deformation in the length of a material under an axial (compressive or tensile) loading, as a change in length per unit length,

$$e = l/L$$
$$= S/E \text{ (Hooke's law)}$$

stress: force per unit area,

$$S = F/A$$

stress, allowable: stress, working ∗

stress, bending: stress on the cross-sectional area of a beam due to the bending moment of the beam under load,

$$S = cM/I = M/Z$$

stress, compressive: stress on the cross-sectional area of a body, normal (90°) to the compression force acting on the body.

stress, shear: stress in the cross-sectional plane of a material that is in the same plane as the shearing force acting on it,

$$S_S = F/A$$

stress, tensile: stress acting on the cross-sectional area of a body, normal (90°) to the tension force acting on the body.

stress, working (S_W): allowable stress considered to be safe in the design of the member. It is the ultimate stress of the material divided by a safety factor,

$$S_W = S_U/\text{sf}$$

tensile strength: ultimate strength (tension) of a material prior to fracture, when subjected to an increasing tension load.

tensile strength (ultimate) of various materials: ultimate strength (tension) of various materials ∗

tensile stress: stress, tensile ∗

torque: moment of a twisting force,

$$T = I_P S/c = SZ_P$$

torsion modulus: modulus of elasticity, shear (G) ∗

STRENGTH OF MATERIALS — *continued*

ultimate strength (S_U): modulus of rupture, maximum stress at point of rupture.

ultimate strength (bending) of various wood materials (1000 lb/in.^{2}):

cedar, red: 6.5	pine, white: 5
cypress: 6	yellow: 7
fir: 5	redwood: 5
hemlock: 5	spruce: 7
oak, white: 8	tamarack: 4.5

ultimate strength (compression) of various materials (1000 lb/in.^{2}):

aluminum, cast: 9	lead, rolled: 7
drawn: 20	limestone: 9
aluminum/bronze: 120	magnesium, extruded: 17
brass, cast: 20	marble: 12
rolled: 25	sandstone: 10
brick: 4	slate: 14
bronze, cast: 55	stainless steel: 225
concrete: 3	steel, cast: 60
copper, cast: 40	cold-rolled: 60
rolled: 32	h.r. high-carbon: 70
copper wire: 38	h.r. low-carbon: 90
granite: 20	structural: 60
iron, cast: 80	tin, cast: 6
malleable: 40	zinc, cast: 18
wrought: 50	

ultimate strength (shear) of various materials (1000 lb/in.^{2}):

aluminum, cast: 11	limestone: 1.4
brass, cast: 36	magnesium, extruded: 17
rolled: 50	marble: 1.3
concrete: 1.2	sandstone: 1.7
copper, cast: 27	steel, cast: 45
granite: 2.3	cold-rolled: 60
iron, cast: 18	h.r. high-carbon: 105
malleable: 48	h.r. low-carbon: 45
wrought: 40	structural: 45

STRENGTH OF MATERIALS — *continued*

ultimate strength (tension) of various materials (1000 lb/in.^{2}):

aluminum, cast: 15
      drawn: 30
      rolled: 26
      wrought: 60
aluminum/bronze: 75
aluminum wire, annealed: 20
brass, cast: 25
      rolled: 60
brass wire: 80
brass wire, annealed: 50
brick: 0.2
bronze, cast: 30
      rolled: 65
bronze/phosphor, cast: 50
bronze/phosphor wire: 100
concrete: 0.3
copper, cast: 25
      rolled: 33
copper wire: 60
copper wire, annealed: 36
german silver: 41
gold, cast: 20
gold wire: 30
granite: 1.2
iron, cast: 20
      malleable: 55
      wrought: 50
lead, cast: 1.8
      rolled: 3.2
limestone: 0.8
magnesium, extruded: 32
      wrought: 35

marble: 1
molybdenum: 75
monel: 75
nickel: 65
phenolic, molded: 7.5
piano wire: 300
platinum wire: 53
platinum wire, annealed: 32
polystyrene: 6
rope, manila: 8
rope, steel: 80
sandstone: 0.3
silver, rolled: 40
slate: 0.4
stainless steel: 225
stainless steel, annealed: 85
steel, annealed: 56
      cast: 60
      cold-rolled: 80
      h.r. high-carbon: 120
      h.r. low-carbon: 60
      structural: 60
steel wire: 120
steel wire, annealed: 80
tin: 2.2
titanium: 72
tungsten: 250
zinc, cast: 5
zirconium: 40

working stress: stress, working   *
yield point: yield strength   *
yield strength: yield point; stress point of a material beyond its elastic limit that produces a definite predetermined deformation of the material.

STRENGTH OF MATERIALS — *continued*

yield strength (tension) of various materials (1000 psi):

aluminum, cast: 9	molybdenum: 55
aluminum/bronze cast: 25	monel: 35
brass, cast: 12	nickel: 28
rolled: 30	stainless steel: 185
bronze, cast: 15	stainless steel, annealed: 32
bronze/phosphorus: 24	steel, annealed: 28
copper: 5.5	cast: 40
iron, cast: 12	tin: 1.3
malleable: 36	titanium: 55
wrought: 38	tungsten: 220
magnesium, extruded: 16	zirconium: 14
wrought: 18	

Young's modulus: modulus of elasticity (E)   *

---

sublimation: transition of a substance directly from the solid phase to the vapor phase without going through the liquid phase, or the reverse of this process.

sublimation heat: HEAT TRANSFER

suction lift: PUMP FORMULAS

surveyor's chain: CHAIN, GUNTER

surveyor's link: LINK, GUNTER

SUS; Saybolt Universal Second: SAYBOLT SECOND UNIVERSAL

symbols of elements: ELEMENTS

synchronous motor: ELECTRICAL

synchronous speed: ELECTRICAL

T

Ta: tantalum (element)
tablespoon =

14.78 cm³   14.78 ml
0.0625 cup   0.5 oz (liquid)
4 dr (liquid)   3 tsp
0.902 in.³

tangent: TRIGONOMETRY
tangential acceleration: ROTATION
tangential velocity: ROTATION
tank, cylindrical: where

C = capacity, gal
D = inside diameter, in.
L = inside length, in.
V = volume, in.³

capacity,

$$C = 0.0034D^2L$$

volume,

$$V = 0.7854D^2L$$

Tb: terbium (element)
Tc: technetium (element)

Te: tellurium (element)

teaspoon =

4.93 cm^3	4.93 ml
1.333 dr (liquid)	0.1667 oz (liquid)
0.30 in.^{3}	0.333 tbsp

TEE SECTION FORMULAS: where

$a, b, c, d,$
$\quad t, x, y$ = dimensions, in.
$\quad\quad A$ = area, in.^{2}
$\quad\quad$ cg = center of gravity
$\quad I_{AA}, I_{BB}$ = moment of inertia, in.^{4}
$\quad k_{AA}, k_{BB}$ = radius of gyration, in.

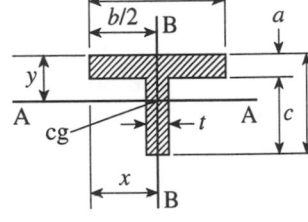

area,

$$A = ab + ct$$

center of gravity (cg) at

$$x = 0.5b$$

$$y = \frac{0.5a^2b + ct(0.5c + a)}{ab + ct}$$

moment of inertia,

$$I_{AA} = 0.333[y^3b + t(d - y)^3 - (b - t)(y - a)^3]$$
$$I_{BB} = 0.0833(b^3a + t^3c)$$

radius of gyration,

$$k_{AA} = (I_{AA}/A)^{1/2}$$
$$k_{BB} = (I_{BB}/A)^{1/2}$$

temperature, absolute: ABSOLUTE TEMPERATURE
temperature conversions: DEGREE C; DEGREE F; DEGREE K; DEGREE R
tensile strength: STRENGTH OF MATERIALS
tera: prefix; equals E + 12
Th: thorium (element)

therm: 100,000 Btu

thermal conductance: HEAT TRANSFER

thermal efficiency: THERMODYNAMICS

thermal expansion, cubical: HEAT TRANSFER

thermal expansion, linear: HEAT TRANSFER

thermistor: device for measuring temperature as a function of electrical resistance.

thermocouple: electrical device for measuring temperature, consisting of two dissimilar metals that generate a voltage at the junction of the two metals, proportional to the change in temperature at the junction.

thermocouple, bimetallic: mechanical device for measuring temperature based on the difference in linear expansion of two dissimilar metals, proportional to the change in temperature.

THERMODYNAMICS: study of transfer of heat energy and related work energy, where

Btu = British thermal unit, unit of heat
C_P = specific heat at constant pressure, Btu/lb/°F
C_V = specific heat at constant volume, Btu/lb/°F
eff = thermal efficiency, decimal
E_K = kinetic energy, ft-lb/lb
E_W = work energy output, ft-lb/lb
h = enthalpy, Btu/lb
k = adiabatic exponent for the gas
$\ln$ = logarithm to base e
n = polytropic exponent for the gas
P = absolute pressure, psia
Q = heat energy, Btu/lb
R = gas constant, ft-lb/lb/°R
s = entropy, Btu/lb/°F or Btu/lb/°R
t = temperature, °F
T = absolute temperature, °R
U = internal energy, Btu/lb
v = specific volume, ft³/lb
V = velocity, ft/sec

THERMODYNAMICS — *continued*

adiabatic compression: external work done on a gas that releases its internal energy and raises its temperature without the addition or loss of heat,

$$144Pv^k = \text{constant}$$

adiabatic expansion: expansion of a gas due to its internal energy, resulting in a drop in gas temperature without addition or loss of heat,

$$144Pv^k = \text{constant}$$

adiabatic exponent for a gas: the ratio of specific heat at constant pressure to specific heat at constant volume.

$$k = C_P/C_V$$

adiabatic exponent (*k*) for various gases (at 60°F and atmospheric pressure).

acetylene: 1.26

air: 1.395

alcohol: 1.13

ammonia: 1.30

argon: 1.68

benzene: 1.08

butane: 1.24

carbon dioxide: 1.30

carbon disulfide: 1.20

carbonic acid: 1.27

carbon monoxide: 1.403

carbon tetrachloride: 1.18

chlorine: 1.34

ethane: 1.21

ethyl chloride: 1.13

ethylene: 1.21

freon F12: 1.14

helium: 1.67

hydrogen: 1.41

hydrogen chloride: 1.41

hydrogen sulfide: 1.30

methane: 1.32

methyl chloride: 1.20

natural gas. 1.269

neon: 1.642

nitric oxide: 1.40

nitrogen: 1.40

nitrous oxide: 1.311

octane: 1.04

oxygen: 1.40

pentane: 1.06

propane: 1.13

steam: 1.33 at 212°F

sulfur dioxide: 1.25

water vapor: 1.33 at 212°F

THERMODYNAMICS — *continued*

adiabatic process, reversible: isentropic process; gas is compressed or expanded without addition or removal of heat, and no change in entropy, as $Pv^k = $ constant.

enthalpy change,

$$h_2 - h_1 = C_P(t_2 - t_1)$$

$$= \frac{kRT_1[(P_2/P_1)^{(k-1)/k} - 1]}{788(k - 1)}$$

entropy change,

$$s_2 - s_1 = 0$$

heat change,

$$Q_2 - Q_1 = 0$$

internal energy change,

$$U_2 - U_1 = 0.185(P_2 v_2 - P_1 v_1)/(k - 1)$$

$$= \frac{RT_1[(T_2/T_1) - 1]}{788(k - 1)}$$

$$= \frac{RT_1[(P_2/P_1)^{(k-1)/k} - 1]}{778(k - 1)}$$

temperature T_2,

$$T_2 = T_1(P_2/P_1)^{(k-1)/k}$$

work energy output,

$$E_W = 144(P_1 V_1 - P_2 V_2)/(k - 1)$$
$$= R(T_1 - T_2)/(k - 1)$$
$$= RT_1[(T_2/T_1) - 1]/(k - 1)$$

constant-pressure process: isobaric process   *
constant-temperature process: isothermal process   *
constant-volume process: isovolume process   *
energy balance equation: steady flow   *

THERMODYNAMICS — *continued*

enthalpy: specific energy value of a material equal to the sum of its internal and flow energies,

$$h = U + 0.185Pv$$

entropy change: term used as a measure of energy that is lost or gained in a process. Entropy is constant (zero change) in an adiabatic, reversible (isentropic) process where heat is not added or lost,

$$s_2 - s_1 = (Q_2 - Q_1)/T$$

internal energy: energy contained in the substance as a result of temperature or pressure changes and cannot be measured in a specific state. However, the changes in internal energy can be measured in various processes.

isentropic process: adiabatic process, reversible   *

isobaric process: process where pressure is held constant, as $v/T = $ constant.

enthalpy change,

$$h_2 - h_1 = C_P(t_2 - t_1)$$
$$= 0.185kP(v_2 - v_1)/(k - 1)$$

entropy change,

$$s_2 - s_1 = C_P \ln (v_2/v_1)$$
$$= C_P \ln (T_2/T_1)$$

heat change,

$$Q_2 - Q_1 = R(k/k - 1)(T_2 - T_1)/778$$
$$= 0.185kP(v_2 - v_1)/(k - 1)$$

internal energy change,

$$U_2 - U_1 = 0.185P(v_2 - v_1)/(k - 1)$$

THERMODYNAMICS — *continued*

(🎱) **isobaric process** — *continued*

temperature T_2,

$$T_2 = T_1(v_2/v_1)$$

work energy output,

$$E_W = 144P(v_2 - v_1)$$

isochoric process: isovolume process   *
isodynamic process: isothermal process   *
isothermal process: isodynamic process; gas is compressed or expanded at constant temperature, as Pv = constant.

enthalpy change,

$$h_2 - h_1 = 0$$

(🎱) entropy change,

$$s_2 - s_1 = R \ln (P_1/P_2)/778$$
$$= R \ln (v_2/v_1)/778$$

(🎱) heat change,

$$Q_2 - Q_1 = RT \ln (P_1/P_2)/778$$
$$= RT \ln (v_2/v_1)/778$$
$$= 0.185P_1v_1 \ln (v_2/v_1)$$

internal energy change,

$$U_2 - U_1 = 0$$

temperature T_2,

$$T_2 = T_1$$

work energy output,

$$E_W = 144P_1v_1 \ln (v_2/v_1)$$
$$= RT_1 \ln (P_1/P_2)$$
$$= RT_1 \ln (v_2/v_1)$$

THERMODYNAMICS — *continued*

isovolume process: isochoric process; process is held at a constant volume, as P/T = constant.

⊗ enthalpy change,

$$h_2 - h_1 = 0.185kv(P_2 - P_1)/(k - 1)$$

⊗ entropy change,

$$s_2 - s_1 = C_V \ln (P_2/P_1)$$
$$= C_V \ln (T_2/T_1)$$

heat change,

$$Q_2 - Q_1 = C_V(t_2 - t_1)$$
$$= 0.185v(P_2 - P_1)/(k - 1)$$
$$= \frac{R(T_2 - T_1)}{778(k - 1)}$$

internal energy change,

$$U_2 - U_1 = 0.185v(P_2 - P_1)/(k - 1)$$

temperature T_2,

$$T_2 = T_1 P_2/P_1$$

work energy output,

$$E_W = 0$$

kinetic energy: energy due to velocity of the gas or fluid,

$$E_K = 0.0155V^2$$

laws of thermodynamics:

first law, conservation of energy: energy can neither be created nor destroyed,

$$Q_1 - Q_2 = U_2 - U_1 + (E_W/778)$$

second law: heat energy that is used to produce work has inherent heat energy losses depending on the efficiency of the system, as the heat energy loss cannot be completely converted into work.

THERMODYNAMICS — *continued*

polytropic process: heat added or removed from the gas in the process is directly proportional to its change in temperature, Pv^n = constant.

(Ⓧ) enthalpy change,

$$h_2 - h_1 = \frac{kRT_1[(P_2/P_1)^{(n-1)/n} - 1]}{778(k-1)}$$

(Ⓧ) entropy change,

$$s_2 - s_1 = C_V[(n-k)/(n-1)] \ln (T_2/T_1)$$
$$= C_V[(n-k)/n] \ln (P_2/P_1)$$

(Ⓧ) heat change,

$$Q_2 - Q_1 = C_V(t_2 - t_1)(n-k)/(n-1)$$
$$= \frac{R(n-k)T_1[(P_2/P_1)^{(n-1)/n} - 1]}{778(k-1)(n-1)}$$

(Ⓧ) internal energy change,

$$U_2 - U_1 = \frac{RT_1[(P_2/P_1)^{(n-1)/n} - 1]}{778(k-1)}$$

work energy output,

$$E_W = 144(P_1v_1 - P_2v_2)/(n-1)$$
$$= RT_1[(P_2/P_1)^{(n-1)/n} - 1]/(n-1)$$

steady flow: energy balance equation, continuous and uniform one-directional flow of a fluid in a system, for a specific time, with no change in velocity, internal energy, specific volume, etc.

$$778U_1 + 144P_1v_1 + E_{K1} + 778Q_1 =$$
$$778U_2 + 144P_2v_2 + E_{K2} + 778Q_2 + E_W$$

thermal efficiency: ratio of work energy output to the heat energy input,

$$\text{eff} = E_W/778Q$$

three phase: ELECTRICAL
Ti: titanium (element)
Tl: thallium (element)
Tm: thulium (element)
ton, ton (short) =

8.896 E + 08 dyn	2000 lb (avdp)
9.07 E + 05 g	32,000 oz (avdp)
14 E + 06 gr	29,166.7 oz (apoth/troy)
20 hwt	0.8929 ton (long)
907.2 kg	0.90718 ton (metric)
2430.6 lb (apoth/troy)	

ton (gross): TON (LONG)
ton (long) =

9.964 E + 08 dyn	2722 lb (apoth/troy)
1.016 E + 06 g	2240 lb (avdp)
15.68 E + 06 gr	3.584 E + 04 oz (avdp)
22.4 hwt	1.12 ton
1016 kg	1.016 ton (metric)

ton (long)/mile =

0.42424 lb/ft
0.03535 lb/in.

ton (metric) =

9.8066 E + 08 dyn	1 millier
1 E + 06 g	9807 N
22.04 hwt	35,273 oz
1000 kg	1.1023 ton
2204.6 lb	0.9842 ton (long)
2679 lb (apoth/troy)	

ton (metric)/kilometer: KILOGRAM/METER
ton (net): TON

ton (refrigeration) =

12,000 Btu/h	3023.95 kg-cal/h
200 Btu/min	37.971 kg of ice melted in 1 h
4.716 hp	83.711 lb of ice melted in 1 h

ton (short): TON
ton/cubic foot =

2000 lb/ft^3	3.2 E$-$04 oz/ft^3
267.36 lb/gal	18.518 oz/in.^{3}
1.1574 lb/in.^{3}	27 ton/yd^3
5.4 E$-$04 lb/yd^3	

ton/cubic yard =

74.074 lb/ft^3	1185.2 oz/ft^3
9.9074 lb/gal	0.68585 oz/in.^{3}
0.04286 lb/in.^{3}	0.03704 ton/ft^3
2000 lb/yd^3	

ton/mile =

0.3788 lb/ft	0.5637 ton (metric)/km
0.03157 lb/in.	

ton/square foot =

0.9451 atm	95.76 kPa
0.9576 bar	2000 lb/ft^2
957,602 baryes	13.889 lb/in.^{2}
71.826 cm (mercury, 0°C, 32°F)	0.71826 m (mercury, 0°C, 32°F)
957,602 dyn/cm^2	718.26 mm (mercury, 0°C, 32°F)
32.0363 ft (water, 4°C, 39.2°F)	95,760 N/m^2
976.485 g/cm^2	95,760 Pa
384.43 in. (water, 4°C, 39.2°F)	0.0069 ton/in.^{2}
0.97648 kg/cm^2	718.26 torr
9764.85 kg/m^2	

ton/square inch =

1.37895 E+08 dyn/cm²	2000 lb/in.²
1.406 E+06 kg/m²	144 ton/ft²

torr: unit of pressure measurement equal to pressure required to support a 1-mm column of mercury at 0°C, 32°F.

torr =

0.0013158 atm	13.5951 kg/m²
0.001333 bar	0.13332 kPa
1333.2 baryes	2.7845 lb/ft²
0.1 cm (mercury, 0°C, 32°F)	0.01934 lb/in.²
1333.22 dyn/cm²	1000 μm (mercury, 0°C, 32°F)
0.0446 ft (water, 4°C, 39.2°F)	1 mm (mercury, 0°C, 32°F)
1.3595 g/cm²	13.595 mm (water, 4°C, 39.2°F)
0.03937 in. (mercury, 0°C, 32°F)	133.32 N/m²
0.5352 in. (water, 4°C, 39.2°F)	133.32 Pa
0.0013595 kg/cm²	0.001392 ton/ft²

torque: ROTATION

torus: RING

trajectory: path of a projectile, projected through space by a force.
TRAJECTORY FORMULAS: where

a = initial angle of projectile, degrees
F = force, lb
g = gravitational acceleration, 32.174 ft/sec²
I = impulse, lb-sec
t = time, sec
t_c = time of force contact with body, sec
V = velocity, ft/sec
V_x = horizontal velocity, ft/sec
V_y = vertical velocity, ft/sec
W = weight of body, lb
x = horizontal distance, ft
y = vertical distance, ft

TRAJECTORY FORMULAS — *continued*

projected by an angular force:

⊗ distance traveled,

$$\text{in } t, \; x = V_x t = Vt \cos a$$
$$\text{maximum distance, } x_{max} = 0.062V^2 (\sin a)(\cos a)$$

⊗ height attained,

$$\text{in } t, \; y = V_y t - 16.087t^2$$
$$= Vt \sin a - 16.087t^2$$
$$\text{maximum height, } y_{max} = 0.0155V^2 \sin^2 a$$

impulse,

$$I = Ft_c$$

time to obtain highest point,

$$t = 0.0311V \sin a$$

time to return to starting level,

$$t = 0.062V \sin a$$

velocity,

$$V = (V_x^2 + V_y^2)^{1/2}$$
$$V_x = V \cos a$$
$$V_y = V \sin a - gt$$

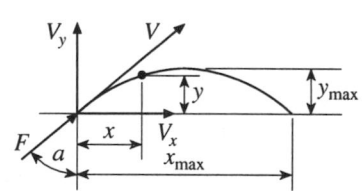

projected by a horizontal force:

direction,

$$\tan a = V_y/V_x$$

⊗ distance traveled in t,

$$x = V_x t$$

⊗ height (drop) in t,

$$y = 16.087t^2$$

TRAJECTORY FORMULAS — *continued*

projected by a horizontal force — *continued*

impulse, $I = Ft_c$
velocity,

$$V = (V_x^2 + V_y^2)^{1/2}$$
$$V_x = gIt/W$$
$$V_y = gt$$

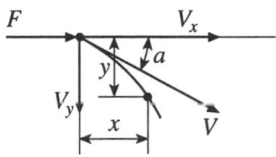

---

transformer: ELECTRICAL

transitional flow: smooth and turbulent flow of a fluid or gas where the calculated Reynold's number is greater than 2000 and less than 4000.

trapezium: four-sided figure with no parallel sides and no equal sides.

area,

$$A = 0.5(ah + by)$$

height,

$$h = c \sin (ABD)$$
$$y = d \sin (CBE)$$

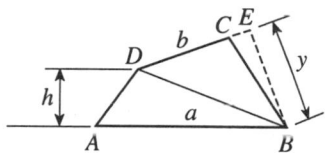

trapezoid: four-sided figure with one pair of opposite sides parallel, as side a is parallel to side b, shown in the figure.

area,

$$A = 0.5h(a + b)$$

center of gravity (cg), on line from x_1 to x_2, at

$$x = 0.333(a^2 + ab + b^2)/(a + b)$$
$$x_1 = 0.5a, \quad x_2 = 0.5b$$
$$y = 0.333h(2a + b)/(a + b)$$
$$z = 0.333h(a + 2b)/(a + b)$$

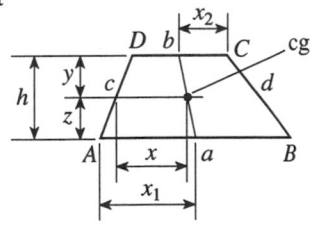

height,

$$h = c \sin A = d \sin B$$

trapezoid, isosceles: four-sided figure with one pair of opposite sides parallel and the nonparallel sides equal.

area,

$$A = 0.5h(a + b)$$
$$= 0.5c(a + b) \sin A$$

height,

$$h = c \sin A$$

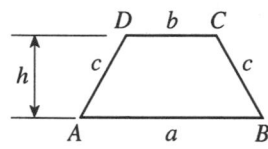

trapezoid section: where

a, b, h, x_1, x_2, y, z = dimensions, in.
A = area, in.^{2}
cg = center of gravity
I_{AA}, I_{BB} = moment of inertia, in.^{4}
k_{AA}, k_{BB} = radius of gyration, in.
Z_{AA}, Z_{BB} = section modulus, in.^{3}

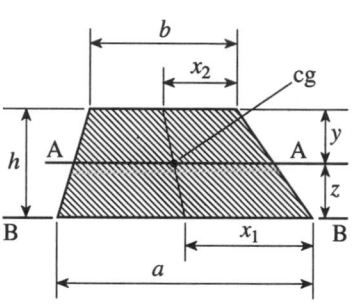

area,

$$A = 0.5h(a + b)$$

center of gravity (cg), on line from x_1 to x_2, at

$$x_1 = 0.5a, \quad x_2 = 0.5b$$
$$y = 0.333h(2a + b)/(a + b)$$
$$z = 0.333h(a + 2b)/(a + b)$$

moment of inertia,

$$I_{AA} = 0.0278h^3(a^2 + 4ab + b^2)/(a + b)$$
$$I_{BB} = 0.0833h^3(a + 3b)$$

radius of gyration,

$$k_{AA} = 0.2357h(a^2 + 4ab + b^2)^{1/2}/(a + b)$$
$$k_{BB} = 0.4082h[(a + 3b)/(a + b)]^{1/2}$$

section modulus,

$$Z_{AA} = 0.0833h^2(a^2 + 4ab + b^2)/(2a + b)$$
$$Z_{BB} = 0.0833h^2(a + 3b)$$

triangle: three-sided figures with the sum of the three included angles equal to 180°.

TRIANGLES: where

A = area
cg = center of gravity
d = diameter of inscribed circle
D = diameter of circumscribed circle
h = height
s = perimeter

equilateral: triangle with three equal sides and three equal angles.

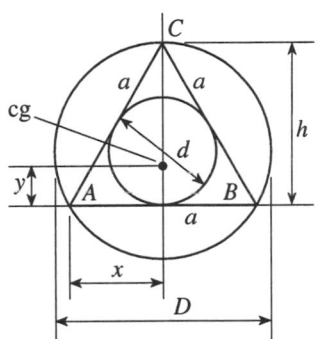

area,

$$A = 0.433a^2 = 0.5ah$$

center of gravity (cg), at

$$x = 0.5a, \quad y = 0.333h$$

height,

$$h = 0.866a$$

diameter of circumscribed circle,

$$D = 1.1547a$$

diameter of inscribed circle,

$$d = 0.5774a$$

perimeter,

$$s = 3a$$

isosceles: triangle with two sides equal and opposite angles equal.

area,

$$A = 0.5bh$$

center of gravity (cg), at

$$x = 0.5b, \quad y = 0.333h$$

TRIANGLES — *continued*

isosceles — *continued*

diameter of circumscribed circle,

$$D = a/\sin A = b/\sin C$$

diameter of inscribed circle,

$$d = b \tan (A/2) = (2a - b) \tan (C/2)$$

height,

$$h = a \sin A$$
$$= [a^2 - (0.5b)^2]^{1/2}$$

perimeter,

$$s = 2a + b$$

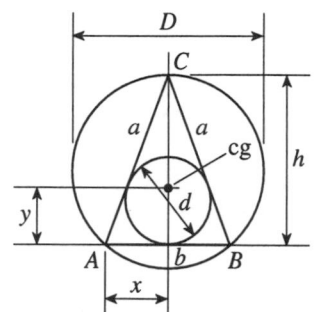

oblique: triangle with unequal sides, unequal angles.

area,

$$A = 0.5bh$$

center of gravity (cg) on line from x to B,

$$x = 0.5b, \quad y = 0.333h$$

diameter of circumscribed circle,

$$D = a/\sin A = b/\sin B = c/\sin C$$

diameter of inscribed circle,

$$d = (s - 2a) \tan (A/2)$$

height,

$$h = a \sin C$$
$$= c \sin A$$

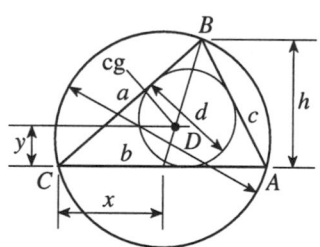

perimeter,

$$s = a + b + c$$

TRIANGLES — *continued*

right: triangle with one (90°) right angle.

area,

$$A = 0.5bh$$

center of gravity (cg) on line from x to B.

$$x = 0.5b, \quad y = 0.333a$$

diameter of circumscribed circle,

$$D = c$$

diameter of inscribed circle,

$$d = a + b - c$$

height,

$$h = a$$

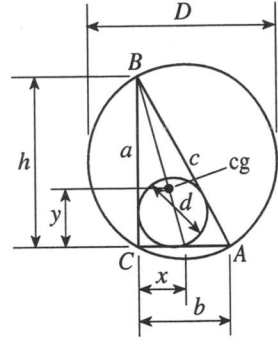

hypotenuse, side opposite the right angle
perimeter,

$$s = a + b + c$$

sides,

$$a = (c^2 - b^2)^{1/2} = c \sin A - b \tan A$$
$$b = (c^2 - a^2)^{1/2} = c \cos A = a/\tan A$$
$$c = (a^2 + b^2)^{1/2} = a/\sin A = b/\cos A$$

TRIANGLE SECTION FORMULAS: where

$a, b, c, x, y,$ = dimensions, in.
A = area, in.^{2}
cg = center of gravity
h = height, in.
I_{AA}, I_{BB}, I_{CC} = moment of inertia, in.^{4}
I_P = moment of inertia, polar, in.^{4}
k_{AA}, k_{BB}, k_{CC} = radius of gyration, in.
Z_{AA}, Z_{BB} = section modulus, in.^{3}
Z_P = section modulus, polar, in.^{3}

TRIANGLE SECTION FORMULAS — *continued*

equilateral triangle:

area,

$$A = 0.433a^2 = 0.5ah$$

center of gravity (cg),

$$x = 0.5a$$
$$y = 0.2887a = 0.333h$$

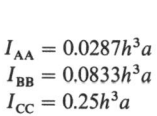

height,

$$h = 0.866a$$

moment of inertia,

$$I_{AA} = 0.0287h^3a$$
$$I_{BB} = 0.0833h^3a$$
$$I_{CC} = 0.25h^3a$$
$$I_P = 0.0361a^4$$

radius of gyration,

$$k_{AA} = 0.2357h$$
$$k_{BB} = 0.4082h$$
$$k_{CC} = 0.707h$$

section modulus,

$$Z_{AA} = 0.0417h^2a$$
$$Z_{BB} = 0.0833h^2a$$
$$Z_{CC} = 0.25h^2a$$
$$Z_P = 0.05a^3$$

oblique triangle:

area,

$$A = 0.5bh$$

center of gravity (cg) on line from x to B,

$$x = 0.5b, \quad y = 0.333h$$

TRIANGLE SECTION FORMULAS—*continued*

oblique triangle—*continued*

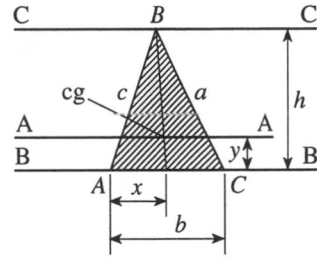

height,

$$h = c \sin A$$

moment of inertia,

$$I_{AA} = 0.0278h^3b$$
$$I_{BB} = 0.0833h^3b$$
$$I_{CC} = 0.25h^3b$$

radius of gyration,

$$k_{AA} = 0.2357h$$
$$k_{BB} = 0.4082h$$
$$k_{CC} = 0.707h$$

section modulus,

$$Z_{AA} = 0.04167h^2b$$
$$Z_{BB} = 0.0833h^2b$$
$$Z_{CC} = 0.25h^2b$$

TRIGONOMETRY: study of the measurement of the triangle and the use of its related functions. *Note:* angle and side solutions are given as one example; other two angles and sides are typical.

acute angle: angle less than 90°.
basic functions of angles:

arc cos A, $\cos^{-1} A =$
    angle of this cos value is A
arc cot A, $\cot^{-1} A =$
    angle of this cot value is A
arc csc A, $\csc^{-1} A =$
    angle of this csc value is A
arc sec A, $\sec^{-1} A =$
    angle of this sec value is A
arc sin A, $\sin^{-1} A =$
    angle of this sin value is A

TRIGONOMETRY — *continued*

basic functions of angles — *continued*

arc tan A, $\tan^{-1} A =$
    angle of this tan value is A
$\cos = \text{cosine} = 1/\sec$
$\cot = \text{ctn} = \text{cotangent} = 1/\tan$
$\csc = \text{cosecant} = 1/\sin$
$\text{ctn} = \text{cotangent} = \cot$
$\sec = \text{secant} = 1/\cos$
$\sin = \text{sine} = 1/\csc$
$\tan = \text{tangent} = 1/\cot$

cosecant: csc

cosine: cos

cosine law: in any triangle, the square of any side is equal to the sum of the squares of the other two sides minus twice the product of those two sides multiplied by the cosine of their included angle, as

$$b^2 = a^2 + c^2 - 2ac \cos B$$
$$\cos B = (a^2 + c^2 - b^2)/2ac$$

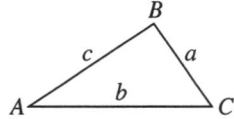

cotangent: cot

formulas:

$$\cos A = b/c = 1/\sec A = \sin A/\tan A$$
$$\cot A = b/a = 1/\tan A = \cos A/\sin A$$
$$= \cos A \csc A$$
$$\csc A = c/a = 1/\sin A$$

double angle,

$$\cos 2A = 2 \cos^2 A - 1$$
$$= \cos^2 A - \sin^2 A$$
$$= 1 - 2 \sin^2 A$$
$$\cot 2A = (\cot^2 A - 1)/2 \cot A$$
$$\sin 2A = 2 \sin A \cos A$$
$$\tan 2A = 2 \tan A/(1 - \tan^2 A)$$
$$= 2/(\cot A - \tan A)$$

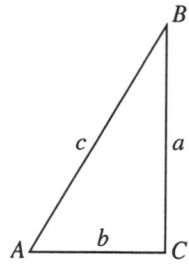

TRIGONOMETRY — *continued*

formulas — *continued*

half angle,

$$\cos (A/2) = 0.707(1 + \cos A)^{1/2}$$
$$\sin (A/2) = 0.707(1 - \cos A)^{1/2}$$
$$\tan (A/2) = \sin A/(1 + \cos A)$$
$$= [(1 - \cos A)/(1 + \cos A)]^{1/2}$$
$$= (1 - \cos A)/\sin A$$

law of cosines: cosine law   *
law of sines: sine law   *
$\sec A = c/b = 1/\cos A$
side: $a = c \sin A = b \tan A$
$\sin A = a/c = 1/\csc A = \cos A/\cot A$
$\sin^2 A + \cos^2 A = 1$
$\tan A = a/b = 1/\cot A = \sin A/\cos A$

obtuse angle: angle greater than 90°

right angle: 90° angle.

secant: sec

sine: sin

sine law: in any triangle, the sides are proportional to the sines of the opposite angles, as

$$a/\sin A = b/\sin B = c/\sin C$$
$$a/b = \sin A/\sin B$$
$$b/c = \sin B/\sin C$$
$$a/c = \sin A/\sin C$$

tangent: tan

trigonometry table of function value (angle in degrees)

angle	sin	cos	tan	angle	sin	cos	tan
0	0.0000	1.000	0.0000	31	0.5150	0.8572	0.6009
1	0.0175	0.9998	0.0175	32	0.5299	0.8480	0.6249
2	0.0349	0.9994	0.0349	33	0.5446	0.8387	0.6494
3	0.0523	0.9986	0.0524	34	0.5592	0.8290	0.6745
4	0.0698	0.9976	0.0699	35	0.5736	0.8192	0.7002
5	0.0872	0.9962	0.0875	36	0.5878	0.8090	0.7265
6	0.1045	0.9945	0.1051	37	0.6018	0.7986	0.7536
7	0.1219	0.9925	0.1228	38	0.6157	0.7880	0.7813
8	0.1392	0.9903	0.1405	39	0.6293	0.7771	0.8098
9	0.1564	0.9877	0.1584	40	0.6428	0.7660	0.8391
10	0.1736	0.9848	0.1763	41	0.6561	0.7547	0.8693
11	0.1908	0.9816	0.1944	42	0.6691	0.7431	0.9004
12	0.2079	0.9781	0.2126	43	0.6820	0.7314	0.9325
13	0.2250	0.9744	0.2309	44	0.6947	0.7193	0.9657
14	0.2419	0.9703	0.2493	45	0.7071	0.7071	1.000
15	0.2588	0.9659	0.2679	46	0.7193	0.6947	1.036
16	0.2756	0.9613	0.2867	47	0.7314	0.6820	1.072
17	0.2924	0.9563	0.3057	48	0.7431	0.6691	1.111
18	0.3090	0.9511	0.3249	49	0.7547	0.6561	1.150
19	0.3256	0.9455	0.3433	50	0.7660	0.6428	1.192
20	0.3420	0.9397	0.3640	51	0.7771	0.6293	1.235
21	0.3584	0.9336	0.3839	52	0.7880	0.6157	1.280
22	0.3746	0.9272	0.4040	53	0.7986	0.6018	1.327
23	0.3907	0.9205	0.4245	54	0.8090	0.5878	1.376
24	0.4067	0.9135	0.4452	55	0.8192	0.5736	1.428
25	0.4226	0.9063	0.4663	56	0.8290	0.5592	1.483
26	0.4384	0.8988	0.4877	57	0.8387	0.5446	1.540
27	0.4540	0.8910	0.5095	58	0.8480	0.5299	1.600
28	0.4695	0.8829	0.5317	59	0.8572	0.5150	1.664
29	0.4848	0.8746	0.5543	60	0.8660	0.5000	1.732
30	0.5000	0.8660	0.5774	61	0.8746	0.4848	1.804

trigonometry table of function value — *continued*

angle	sin	cos	tan	angle	sin	cos	tan
62	0.8829	0.4695	1.881	77	0.9744	0.2250	4.332
63	0.8910	0.4540	1.963	78	0.9781	0.2079	4.705
64	0.8988	0.4384	2.050	79	0.9816	0.1908	5.145
65	0.9063	0.4226	2.146	80	0.9848	0.1736	5.671
66	0.9135	0.4067	2.246	81	0.9877	0.1564	6.314
67	0.9205	0.3907	2.356	82	0.9903	0.1392	7.115
68	0.9272	0.3746	2.475	83	0.9925	0.1219	8.144
69	0.9336	0.3584	2.605	84	0.9945	0.1045	9.514
70	0.9397	0.3420	2.748	85	0.9962	0.0872	11.43
71	0.9455	0.3256	2.904	86	0.9976	0.0698	14.30
72	0.9511	0.3090	3.078	87	0.9986	0.0523	19.08
73	0.9563	0.2924	3.271	88	0.9994	0.0349	28.64
74	0.9613	0.2756	3.487	89	0.9998	0.0175	57.29
75	0.9659	0.2588	3.732	90	1.000	0.0000	∞
76	0.9703	0.2419	4.011				

troy weight: unit of measure used in the weighing of gold, silver, jewels, etc. The basis of the weight system uses 12 oz/lb and 20 pwt/oz.

true power: ELECTRICAL

turbulent flow: when a gas or fluid flows through a pipe or duct in an irregular motion with considerable disturbance or mixing. In this case, the calculated Reynold's number is greater than 4000.

two phase: ELECTRICAL

U

U: uranium (element)
ultimate strength: STRENGTH OF MATERIALS
unit conversion: CONVERSION OF UNITS
universal gas constant: GAS CONSTANT, UNIVERSAL

V

V: vanadium (element)

vacuum, perfect: complete vacuum, equal to 0 in. (mercury).

vapor density: weight of 1 ft^3 of a vapor at standard conditions, 70°F temperature and 14.696 lb/in.^{2} pressure.

vapor pressure: absolute pressure of a liquid and its vapor at a specific temperature, in a closed container, where the vapor maintains equilibrium with the liquid.

vapor pressure of various liquids psia at (°F):

acetone: 7.6 (100)
alcohol: 0.86 (68)
benzene: 1.3 (100)
benzol: 3.2 (100)
carbon disulfide: 11.1 (100)
ether: 8.37 (68)
ethyl ether: 11.48 (83)
          16.8 (100)
gasoline: 11.5 (100)
mercury: 2.51 E−05 (68)

methyl acetate: 7.1 (100)
methyl ethyl ketone: 3.5 (100)
naphtha: 0.5 (100)
naphtha (VMP): 1.0 (100)
toluene: 1.0 (100)
toluol: 1.0 (100)
water: 0.08896 (32)
      0.33845 (68)
      0.609 (86)
     14.696 (212)

vaporization heat: HEAT TRANSFER

varas = 0.848176 m

velocity: rate of movement of an object; distance per unit time

velocity, angular: ROTATION

velocity, linear: LINEAR MOTION
velocity of light, light speed =

2.99776 E + 10 cm/s   2.99776 E + 08 m/s
5.9 E + 10 ft/min   6.706 E + 08 mi/h
9.835 E + 08 ft/s   1.1176 E + 07 mi/min
1.180 E + 10 in./s   186,000 mi/s
8.3271 E + 04 m/h

velocity of sound, sonic speed =

3.3162 E + 04 cm/s   331.62 m/s
6.5280 E + 04 ft/min   741.8 mi/h
1088 ft/s   12.36 mi/min
1.3056 E + 04 in./s   0.206 mi/s
1.194 E + 06 m/h

velocity pressure: AIR FLOW; FLUID FLOW; GAS FLOW
vena contracta: FLUID FLOW
venturi meter: FLUID FLOW
VIBRATION: the repeating back-and-forth motion of a body after an
     initial force displaced the body and then the force was removed,
     where

d_F = deflection (force), in.
d_S = deflection (static), in.
f = frequency of vibration, c/s
F = force, lb
k = spring constant, lb/in.
t = time, s/c
W = weight, lb

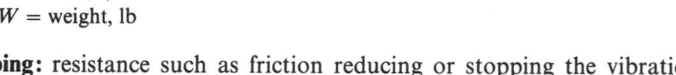

     damping: resistance such as friction reducing or stopping the vibration of
         the body.
     deflection (force): initial displacement due to a force temporarily applied to
         the body,

$$d_F = F/k$$

VIBRATION — *continued*

deflection (static): normal displacement due to the weight of the body in equilibrium,

$$d_S = W/k$$

frequency of vibrations:

$$f = 0.90276/d^{1/2}$$
$$= 5.67(k/W)^{1/2}$$

spring constant: $k = F/d_F$

---

viscosity: internal resistance of the flow of a fluid or gas.

viscosity, absolute; absolute viscosity; coefficient of viscosity; dynamic viscosity: internal friction of a fluid determined in a laboratory by the force required to rotate a disk in a specific fluid using the Brookfield viscometer instrument. Absolute viscosity is the ratio of shearing stress to the rate of shear. Absolute viscosity is the product of kinematic viscosity times density.

English system of units:

lb (force)-s/ft^2	lb (mass)/in./s
lb (force)-s/in.^{2}	pdl-s/ft^2
lb (mass)/ft/h	slug/ft/h
lb (mass)/ft/min	slug/ft/s
lb (mass)/ft/s	slug/in./h
lb (mass)/in./h	

Metric system of units:

cP	μP
dyne-s/cm^2	mPa-s
g/cm/s	N-s/m^2
kg/m/h	P
kg/m/s	Pa-s
kg-s/m^2	

viscosity (absolute) of various gases: GAS FLOW FORMULAS

viscosity (absolute) of various liquids: FLUID FLOW
viscosity, coefficient of: VISCOSITY, ABSOLUTE
viscosity, dynamic: VISCOSITY, ABSOLUTE
viscosity, kinematic; kinematic viscosity: measure of a timed gravity
flow of a fluid down through a calibrated tube at a constant temperature as determined in a laboratory. Kinematic viscosity is the
ratio of absolute viscosity/mass density.

English system of units:

ft^2/h	$in.^2/s$
ft^2/s	Saybolt Second Furol
$in.^2/h$	Saybolt Universal Seconds

Metric system of units:

cm^2/s	m^2/s
cSt	mm^2/s
m^2/h	St

viscosity (kinematic) of various gases: GAS FLOW FORMULAS
viscosity (kinematic) of various liquids: FLUID FLOW
viscosity, specific: FLUID FLOW
volt: ELECTRICAL
volt =

1 E+08 abvolts	1000 mV
1 J/C	0.003333 statvolt
0.001 kV	

volt/centimeter =

1 E+08 abvolts/cm	0.008473 statvolts/in.
0.003333 statvolts/cm	

volt/inch =

3.937 E+07 abvolts/cm	0.001312 statvolt/cm
3.937 E−04 kV/cm	

volt/meter = 1 E+06 abvolt/cm

volt/mil =

> 0.3937 kV/cm
> 1.31362 statvolts/cm

voltage: ELECTRICAL
voltage drop: ELECTRICAL
volt-ampere: ELECTRICAL
volt-coulomb: JOULE
volt-second: WEBER
volume, specific; specific volume: volume of a substance per unit weight and it is equal to the reciprocal of its specific weight.

W

W: tungsten (element)
WATER:

boiling point = 100°C, 212°F
density: specific weight   *
freezing point = 0°C, 32°F
gal = 8.33 lb
heat of fusion = 144 Btu/lb
heat of vaporization = 970 Btu/lb
heat required to evaporate 1 lb at 212°F =

970.89 Btu	0.2844 kWh
1.0237 E + 06 J	284.4 Wh

heat required to raise temperature of 1 lb from 62 to 212°F =

150.02 Btu	0.044 kWh
1.58 E + 05 J	44 Wh

melting point = 0°C, 32°F
specific gravity = 1.0
specific heat = 1 Btu/lb/°F
specific volume, volume =

1.000 cm³/g at 0°C, 32°F
0.01602 ft³/lb at 0°C, 32°F

WATER — *continued*

specific weight, density, weight =

1 g/cm³ at 0°C, 32°F   8.3457 lb/gal at 0°C, 32°F
62.43 lb/ft³ at 0°C, 32°F   8.33 lb/gal at 20°C, 68°F
62.317 lb/ft³ at 20°C, 68°F

viscosity (absolute) at 68.4°F = 1 cP
volume: specific volume   *
weight: specific weight   *

---

watt: ELECTRICAL

watt =

3.414 Btu/h   0.001341 hp
0.05692 Btu/min   1.0194 E − 04 hp (boiler)
9.484 E − 04 Btu/s   0.00136 hp (metric)
1 E + 07 dyn-cm/s   0.01 hW
1 E + 07 erg/s   1 J/s
2655 ft-lb/h   0.860 kg-cal/h
44.26 ft-lb/min   0.01434 kg-cal/min
0.7376 ft-lb/s   2.39 E − 04 kg-cal/s
1424 ft-pdl/min   0.10197 kg-m/s
860.38 g-cal/h   0.001 kW
14.34 g-cal/min   35.52 l-atm/h
0.239 g-cal/s   1 E − 06 MW
10,197 g-cm/s

watt/square centimeter =

3172 Btu/h/ft²   860.38 g-cal/h/cm²
41,113 ft-lb/min/ft²

watt/square inch =

491.65 Btu/h/ft²   6372.54 ft-lb/min/ft²
8.194 Btu/min/ft²   133.39 g-cal/h/cm²

watt-hour: ELECTRICAL
watt-hour =

3.413 Btu	0.00134 hp-h
3.5525 E + 04 cm³-atm	3.1873 E + 04 in.-lb
3.60 E + 10 erg	3600 J
2655 ft-lb	0.860 kg-cal
860 g-cal	367.1 kg-m
3.671 E + 07 g-cm	0.001 kWh

watt-hour-meter/hour-square meter-K =

6.9396 Btu-in./h-ft²-°F

watt-second: electrical energy unit
watt-second: JOULE
watt-second-centimeter/second-square centimeter-°C =

694 Btu-in./h-ft²-°F

wave frequencies of various waves (Hz):

audio: 440
carrier currents: 3 E + 04
cosmic: 3 E + 23
dielectric heating: 5 E + 06
electric AC power: 60
gamma: 1 E + 21
induction heating: 1 E + 04

infrared heating: 1E + 14
radar: 10 E + 10
radio: 1 E + 06
solar radiation: 7.1 E + 14
television: 1 E + 07
ultrasonic: 2 E + 05
X-rays: 1 E + 20

wavelength ranges of various rays (µm):

audio: 7 E + 11 to 1 E + 10
cosmic: 1 E − 06 to 1 E − 12
dielectric heating: 50
electric AC power: 5 E + 12
gamma: 1 E − 06 to 140 E − 06
heat: 0.8 to 400
induction heating: 3 E + 10
infrared: 0.8 to 400
light: 0.4 to 0.8

microwave: 3 E + 04
radar: 3 E + 04
radio: 10 E + 06 to 30 E + 09
solar radiation: 0.42
television: 3 E + 07
ultrasonic: 1.3 E + 09
ultraviolet: 0.014 to 0.4
X-rays: 6 E − 06 to 0.1

weber: metric unit for magnetic flux.
weber =

1 E+05 kilolines	0.003336 statweber
1 E+08 lines	1 V-s
1 E+08 Mx	

weber/square centimeter: metric unit for magnetic flux density.
weber/square centimeter =

1 E+08 G	1 E+08 Mx/cm^2
1 E+08 $lines/cm^2$	6.4516 $Wb/in.^2$
6.4516 E+08 $lines/in.^2$	1 E+04 Wb/m^2

weber/square inch =

1.55 E+07 G	1 E+08 $Mx/in.^2$
1.55 E+07 $lines/cm^2$	0.155 Wb/cm^2
1 E+08 $lines/in.^2$	1550 Wb/m^2
1.55 E+07 Mx/cm^2	

weber/square meter: metric unit for magnetic flux density.
weber/square meter =

1.256 E−06 A turns/m	1 E+04 Mx/cm^2
1 E+04 G	1 E−04 Wb/cm^2
64,516 $lines/in.^2$	6.452 E−04 $Wb/in.^2$

weight, atomic: ATOMIC WEIGHT
weight of a body: the force of gravitational attraction that the Earth
exerts on the body, depending on the mass of the body, mass of the
Earth, and the distance of the body from the center of the Earth. It
is proportional but not identical to the mass of the body, where

g = gravitational acceleration, 32.174 ft/s^2
m = mass of body, slugs
w = weight of body, lb

weight,

$$w = mg$$

weight of copper wire: COPPER WIRE GAUGES

weight, specific: specific weight of a material per unit volume and it is equal to the reciprocal of its specific volume,

English units, lb/ft^3
metric units, g/cm^3

weight (specific) of various gases and vapors (lb/ft^3 at 70°F and atmospheric pressure):

acetic acid: 0.1554
acetone: 0.1503
acetylene: 0.0673
air: 0.075
alcohol: 0.1200
ammonia: 0.0444
argon: 0.1033
benzene: 0.2020
benzine: 0.1875
benzol: 0.2020
butane: 0.155
butylene: 0.1455
camphor: 0.3967
carbon dioxide: 0.1147
carbon disulfide: 0.1973
carbon monoxide: 0.0725
carbon tetrachloride: 0.3981
chlorine: 0.1835
chloroform: 0.3094
ethane: 0.0778
ethanol: 0.1192
ether: 0.1924
ethyl alcohol: 0.1192
ethyl chloride: 0.1774
ethylene: 0.0726
fluorine: 0.0997
formaldehyde: 0.078

formic acid: 0.1191
freon F12: 0.3130
    F22: 0.2383
gasoline: 0.2625
glycerine: 0.2383
helium: 0.0103
heptane: 0.2593
hexane: 0.2182
hydrochloric acid: 0.0945
hydrocyanic acid: 0.0699
hydrofluoric acid: 0.1778
hydrogen: 0.0052
hydrogen chloride: 0.0944
hydrogen fluoride: 0.0518
hydrogen sulfide: 0.0882
kerosene: 0.3375
krypton: 0.2167
manufacturer's gas: 0.0337
mercury vapor: 0.520
methane: 0.0415
methanol: 0.0829
methyl alcohol: 0.0829
methyl chloride: 0.1339
methyl ether: 0.1192
methyl ethyl ketone: 0.1866
mineral spirits: 0.2924
naphthalene: 0.3317

weight (specific) of various gases and vapors — *continued*

naphtha VMP: 0.2812
natural gas: 0.0504
neon: 0.0522
nicotine: 0.420
nitric oxide: 0.0778
nitrogen: 0.0725
nitrogen oxide: 0.0779
nitroglycerine: 0.5878
nitrous oxide: 0.1139
octane: 0.2956
oxygen: 0.0829
phenol: 0.2434
phosphorus: 0.3214
producer's gas: 0.0645

propane: 0.114
propylene: 0.1089
steam (212°F): 0.0466
styrene: 0.2695
sulfur dioxide: 0.1658
toluene: 0.2385
toluol: 0.2325
trichloroethylene: 0.3401
turpentine: 0.3526
water vapor (212°F): 0.0466
wood alcohol: 0.0829
xenon: 0.3398
xylene: 0.2745
xylol: 0.2775

weight (specific) of various liquids (lb/gal):

acetaldehyde: 6.48
acetic acid: 8.75
acetone: 6.55
alcohol: 7.33
ammonia: 6.87
asphalt: 8.29
benzene, benzol: 7.30
benzine: 6.16
brine (25%): 10.0
bromine: 26.07
butyric acid: 8.0
carbolic acid: 8.91
carbon disulfide: 10.5
carbon tetrachloride: 13.24
castor oil: 8.0
coconut oil: 7.5
cottonseed oil: 7.75
diesel oil: 7.5
ethyl acetate: 7.41

ethyl alcohol, ethanol: 6.58
ethylene glycol: 9.25
formaldehyde: 8.96
formic acid: 10
fuel oil, no. 1: 6.91
    no. 2: 7.04
    no. 3: 7.27
    no. 4: 7.50
    no. 5: 7.85
    no. 6: 8.22
gasoline: 5.83
glue, 66% water: 9.16
glycerine: 10.5
grain alcohol: 6.58
hydraulic oil: 7.08
hydrochloric acid: 9.66
hydrocyanic acid: 5.75
hydrofluoric acid: 9.66
hydrogen peroxide: 12.16

weight (specific) of various liquids — *continued*

inerteen: 12.9
isopropyl alcohol: 6.66
kerosene: 6.83
lard oil: 7.62
linseed oil: 7.75
mercury: 113.3
methyl acetate: 7.75
methyl alcohol, methanol: 6.56
methylene chloride: 10.5
methyl ethyl ketone: 6.71
milk: 8.58
mineral oil: 7.58
mineral spirits: 6.66
molasses: 11.66
muriatic acid: 10
naphtha: 7.08
naphtha, VMP: 6.25
nitric acid: 12.5
oil, crude: 7.5
    heavy lubricating: 7.75
    light lubricating: 7.08
olive oil: 7.58

paint: 12
palm oil: 8.00
paraffin: 7.5
petroleum: 7.25
sea water: 8.58
shellac: 10
sodium silicate: 12
sulfuric acid: 15.2
tallow: 7.46
toluol: 7.48
trichloroethylene: 12.2
tung oil: 7.83
turpentine: 7.25
varnish: 6.91
vegetable oil: 7.75
vinegar: 9.0
water: 8.33
whale oil: 7.66
wood alcohol: 6.58
xylene: 7.5
xylol: 7.24

weight (specific) of various materials (lb/ft^3):

acetaldehyde: 48.5
acetic acid: 65.4
acetone: 49
alabaster: 168
alcohol: 55
aluminum: 168
amber: 68
americium: 729
ammonia: 51
antimony: 413
arsenic: 357

asbestos: 153
ashes, cinder: 40
asphalt: 62
babbitt, tin type: 461
bakelite: 84
barium: 234
barytes, loose: 144
    solid: 265
bauxite, loose: 70
    solid: 159
beeswax: 60

weight (specific) of various materials — *continued*

benzene, benzol: 55
benzine: 46
beryllium: 113
bismuth: 610
borax: 109
boron: 159
brass: 530
brick, common: 112
    fire: 137
    hard: 125
    paving: 150
    rubble: 140
brine (25% NaCl): 75
bromine: 195
bronze, aluminum: 480
    phosphor: 553
butyric acid: 60
cadmium: 539
calcium: 97
calcium carbonate: 126
calcium chloride: 134
camphor: 62.3
camphor oil: 55
carbide: 50
carbolic acid: 67
carbon: 138
carbon disulfide: 79
carbon tetrachloride: 99
carborundum: 140
castor oil: 60
cement, portland, loose: 92
        packed: 110
        solid: 193
ceramics: 156
cerium: 430

cesium: 118
chalk: 156
charcoal, oak: 33
    pine: 25
chromium: 448
cinders, coal: 40
clay, damp: 110
    dry: 62.3
coal, solid, anthracite: 95
        bituminous: 79
        lignite: 78
cobalt: 554
coconut oil: 56
coke, solid: 84
columbium: 452
concrete: 150
constantan: 554.6
copper: 555.8
copper ore: 261.7
copper oxide: 190
cork: 16
corkboard: 10
corn oil: 57
cotton flax: 92.8
cottonseed oil: 58
creosote: 67
diamond: 219
diesel oil: 56
dolomite: 181
duralumin: 174
earth: 96
emery: 249
erbium: 572
ethyl acetate: 55
ethyl alcohol, ethanol: 49

weight (specific) of various materials — *continued*

ethyl bromide: 89
ethylene chloride: 80
ethylene glycol: 69
europium: 572
feldspar: 160
felt: 13.7
fiberboard: 15
fiberglass: 4
flaxseed: 45
flint: 162
flour, loose: 28
      packed: 47
fly ash: 75
formaldehyde: 67
formic acid: 75
freon F11: 92
     F12: 82
     F22: 75
fuel oil no. 1: 52
     no. 2: 53
     no. 3: 55
     no. 4: 56
     no. 5: 59
     no. 6: 62
Fuller's earth: 30
gadolinium: 496
galena: 465
gallium: 369
gasoline: 44
germanium: 333
German silver: 525
glass: 156
glue (66% water): 69
glycerine: 79
gold: 1205

grain: 48
grain alcohol: 49
granite: 164
graphite: 131
gravel: 110
gum, Arabic: 90
gypsum: 143
hafnium: 810
heptane: 42
hexane: 41
holmium: 629
hydraulic oil: 53
hydrochloric acid: 72
hydrocyanic acid: 43
hydrofluoric acid: 72
hydrogen chloride: 79
hydrogen peroxide: 91
ice: 57
inconel: 533
indium: 455
inerteen: 97
invar (64 Fe/36 Ni): 497
iodine: 307
iridium: 1400
iron, cast: 450
     malleable: 456
     pure: 491
     wrought: 485
iron ore, hematite: 327
     limonite: 237
     magnetite: 315
iron slag: 171
isopropyl alcohol: 50
ivory: 115
jet fuel: 50

weight (specific) of various materials — *continued*

kerosene: 51
lanthanum: 384
lard oil: 57
lead: 710
lead ore, galena: 465
leather: 59
lignite: 78
limestone, solid: 162
linseed oil: 58
lithium: 33
lutethium: 608
lye (33% water): 106
magnesia, 85%: 19
magnesium: 108
malt, dry: 35
manganese: 464
manganese ore, pyrolusite: 259
manganin: 531
marble: 169
mercury: 847
methyl acetate: 58
methyl alcohol, methanol: 49
methyl chloride: 57
methylene chloride: 79
methyl ethyl ketone: 52
mica: 177
micarta: 85
milk: 64
mineral oil: 57
mineral spirits: 50
mineral wool: 10
molasses: 87
molybdenum: 636
monel (67% Ni): 553
mortar: 94

muriatic acid: 75
naphtha: 53
naphtha VMP: 47
Neat's foot oil: 57
neodymium: 440
nichrome (80 Ni/20 Cr): 523
nickel: 555
nicotine: 63
niobium: 536
nitric acid: 94
nitroglycerine: 100
octane: 44
oil, crude: 56
    lubricating, heavy: 58
    lubricating, light: 53
oleic acid: 56
olive oil: 57
osmium: 1408
oxalic acid: 104
paint: 90
palladium: 748
palm oil: 58
paper: 58
paraffin: 56
peanut oil: 57
peat: 50
pentane: 39
petroleum: 54
phenol: 67
phenolic, molded: 84
phosphor bronze: 548
phosphoric acid: 111
phosphorus, red: 143
            yellow: 114
piano wire: 488

weight (specific) of various materials — *continued*

pine oil: 54
pitch: 70
plaster: 106
plaster of Paris: 137
platinum: 1333
plexiglas: 74
plutonium: 1228
polonium: 577
polyethylene: 57
polystyrene, molded: 67
porcelain: 150
potash: 57
potassium: 54
potatoes: 42
praseodymium: 414
propane: 31.5
propyl alcohol: 50
propylene: 32
Prussic acid: 43
pumice stone: 56
pyralin: 84
pyrite, copper ore: 262
quartz: 162
quenching oil: 56.7
radium: 312
rapeseed oil: 57
resin: 67
rhenium: 1277
rhodium: 775
rock salt: 135
rockwool: 17
rubber, hard: 76
      soft: 57
rubidium: 95
ruthenium: 760
salicylic acid: 90

salt: 131
sand, loose: 95
      packed: 120
sandstone: 143
sawdust: 25
scandium: 156
sea water: 64
selenium: 300
shale: 162
shellac: 75
silica: 180
silicon: 145
silver: 655
slag, crushed: 80
      furnace: 60
      ground: 23
slate, solid: 174
snow, loose: 8
      wet: 50
soap powder: 22
soapstone: 168
soda ash: 30
sodium: 60
sodium carbonate: 151
sodium chloride: 135
sodium hydroxide: 64
sodium nitrate: 140
sodium silicate: 90
solder (50 Pb/50 Sn): 580
soybean oil: 58
starch: 95
stearic acid: 53
steel, cast: 487
      silicon: 477
      stainless: 494
      structural: 485

weight (specific) of various materials — *continued*

Stoddard solvent: 50
stone, crushed: 100
    solid: 150
strontium: 162
styrene: 57
sugar: 105
sulfur: 125
sulfuric acid: 114
talc: 168
tallow: 57
tantalum: 1035
tar: 62
tartaric acid: 104
technetium: 716
tellurium: 389
terbium: 520
thallium: 740
thorium: 710
thulium: 584
tile: 75
tin: 455
tin ore, cassiterite: 417
titanium: 280
toluene: 54
toluol: 56
trichloroethylene: 91
tung oil: 58
tungsten: 1203
turpentine: 54
uranium: 1165
urea: 93
vanadium: 368
varnish: 52
vegetable oil: 58
vinegar: 67
water: 62.317

whale oil: 57
wood: arborvitae: 18
    ash: 42
    balsam: 25
    bamboo: 24
    basswood: 22
    beech: 47
    birch: 40
    butternut: 26
    cedar: 23
    cherry: 40
    chestnut: 37
    cottonwood: 28
    cypress: 30
    ebony: 77
    elm: 36
    fir, Douglas: 32
    gum: 34
    hemlock: 28
    hickory: 50
    locust: 45
    magnolia: 34
    mahogany: 35
    maple, red: 44
        sugar: 38
        white: 33
    oak, red: 42
        white: 48
    pine, white: 27
        yellow: 34
    poplar: 30
    redwood: 26
    spruce: 25
    sycamore: 35
    tamarack: 31
    tupelo: 32
    walnut: 38
    willow: 30
wood alcohol: 49

weight (specific) of various materials — *continued*

wool: 81.5
xylene: 56
xylol: 54
ytterbium: 436
yttrium: 344
zinc: 445

zinc chloride: 181
zinc ore, blende: 248
zinc oxide: 349
zircon: 293
zirconium: 405

weir: FLUID FLOW

wet bulb temperature: maximum temperature reading of an air flow, indicated by a thermometer with its bulb covered by a film of water (as with a water-saturated cloth) and exposed to the stream of air flow.

wire, aluminum: ALUMINUM WIRE GAUGES

wire, copper: COPPER WIRE GAUGES

wire, steel: STEEL WIRE GAUGES

wire mesh: MESH

wood alcohol: methanol; methyl alcohol

working stress: STRENGTH OF MATERIALS

X

Xe: xenon (element)

Y

Y: yttrium (element)

yard: unit of length measurement; also used as meaning "cubic yard" in construction and excavating industries.

yard =

0.04545 chain (Gunter)	36 in.
0.03 chain (Ramden)	9.144 E−04 km
91.44 cm	4.9383 E−04 kn
2 cubits	0.9144 m
0.09144 dam	5.6818 E−04 mi
0.5 fathom	35,997 mil
3 ft	914.4 mm
0.004546 furlong	0.181818 rod
0.009144 hm	

Yb: ytterbium (element)

year =

365 d	12 months
8760 h	52 weeks

yield strength: STRENGTH OF MATERIALS

Young's modulus of elasticity: STRENGTH OF MATERIALS

Z

Z BAR, SECTION FORMULAS: where

a, b, c, d, e, t = dimensions, in.
A = area, in.^{2}
cg = center of gravity
I_{AA}, I_{BB} = moment of inertia, in.^{4}
k_{AA}, k_{BB} = radius of gyration, in.
Z_{AA}, Z_{BB} = section modulus, in.^{3}

area:

$$A = t(2b + e)$$

center of gravity (cg):

$$x = 0.5t$$
$$y = 0.5d$$

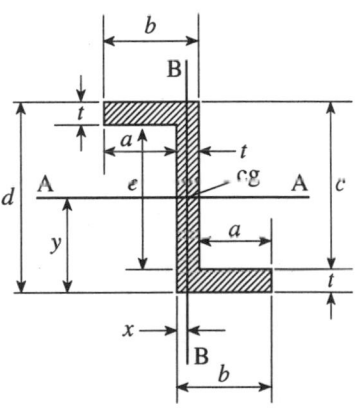

moment of inertia:

$$I_{AA} = 0.0833[d^3b - a(d - 2t)^3]$$
$$I_{BB} = 0.0833[d(b + a)^3 - 2ca^3 - 6acb^2]$$

radius of gyration:

$$k_{AA} = \frac{0.2887[d^3b - a(d - 2t)^3]^{1/2}}{(2bt + et)^{1/2}}$$

$$k_{BB} = \frac{0.2887[d(b + a)^3 - 2ca^3 - 6acb^2]^{1/2}}{(2bt + et)^{1/2}}$$

Z BAR, SECTION FORMULAS — *continued*

section modulus:

$$Z_{AA} = [d^3 b - a(d - 2t)^3]/6d$$

$$Z_{BB} = \frac{[d(b + a)^3 - 2ca^3 - 6acb^2]}{b - 0.5t}$$

Zn: zinc (element)
Zr: zirconium (element)

INDEX

G

H

LICENSE AND LIMITED WARRANTY AGREEMENT

Engineering Formulas Interactive

INSTALLATION PROCEDURE

Engineering Formulas Interactive can be installed on any Windows PC with a CD-ROM drive. After installing EF Interactive, the program can be started from Windows 95, Windows 98, Windows NT start menu or from the EF Interactive icon installed in the Windows 3.1X program manager.

Windows 95, 98, or NT

1. Insert CD in CD-ROM drive.
2. Choose Run... from the Start Menu.
3. In the Open box, type d:/setup, where "d" is the letter assigned to the CD-ROM drive.
4. Click OK, and follow the screen instructions.

Windows 3.1X

1. Insert CD in CD-ROM drive.
2. In the Windows Program Manager, choose Run... from the File Menu.
3. In the Command Line box, type d:/setup, where "d" is the letter assigned to the CD-ROM drive.
4. Click OK, and follow the screen instructions.

USER'S GUIDE

The Help Function contains a **User's Guide** that explains how to: Work with Equation Sets; Work with Equations; Explore & Animate Data; and Work with Graphs. When you click on a formula in the Library, the equation description and parameter definitions are displayed on the screen. The **Formula Pad** then allows you to evaluate the interactive formulas. Use the button bar at the top of the **Formula Pad** to perform the following activities: Click **Solve** to start the evaluation process; Click **Library** for a list of all the equations for which formula pads exist; Click **Copy** to copy the contents of the **Formula Pad** to the clipboard; Click **Intellisim** to launch the full Intellisim equation-solving software.

Minimum System Requirements: Windows 3.1/ 95 / 98 / NT
CD-ROM drive
VGA compatible monitor
16 Mb available RAM
10 Mb available HD space